Lecture Notes in Mathematics 2117

W0042027

More information about this series at
http://www.springer.com/series/304

Stefan Liebscher

Bifurcation
without Parameters

 Springer

Stefan Liebscher
Institur für Mathematik
Freie Universität Berlin
Berlin
Germany

ISBN 978-3-319-10776-9   ISBN 978-3-319-10777-6 (eBook)
DOI 10.1007/978-3-319-10777-6
Springer Cham Heidelberg New York Dordrecht London

Lecture Notes in Mathematics ISSN print edition: 0075-8434
                             ISSN electronic edition: 1617-9692

Library of Congress Control Number: 2014954087

Mathematics Subject Classification (2010): 34C23, 34C20, 34C37, 37G99, 35B32

Printed on acid-free paper

Springer is part of Springer Science+Business Media (www.springer.com)

Preface

This monograph is devoted to the study of dynamical systems with manifolds of equilibria near points at which normal hyperbolicity of these manifolds is violated. It is targeted at mathematicians with at least basic familiarity with classical bifurcation theory. Although methods and concepts are briefly introduced, prior knowledge of center-manifold reductions and normal-form calculations may help to appreciate the presentation.

Manifolds of equilibria arise frequently in parameter dependent systems—by continuation of a trivial equilibrium. Loss of hyperbolicity of such equilibria yields qualitative changes of the local dynamics. Its study is one of the main objectives of classical bifurcation theory.

Here, however, we are interested in manifolds of equilibria that are not caused by additional parameters. Still, qualitative changes of the local dynamics close to the manifold of equilibria occur at points at which normal hyperbolicity of these manifolds breaks down. To exclude not only given but also any unknown or "hidden" parameters, we require the absence of any flow-invariant foliation transverse to the manifold of equilibria at the bifurcation point. We call the emerging theory *bifurcation without parameters*.

On first glance our setting appears to be very degenerate. Indeed, vector fields with manifolds of equilibria form a set of infinite codimension in the space of all smooth vector fields. However, there is a surprisingly rich and diverse collection of applications ranging from networks of coupled oscillators, viscous and inviscid profiles of stiff hyperbolic balance laws, standing waves in fluids, binary oscillations in numerical discretizations, population dynamics, memristor circuits, cosmological models, and many more.

Note that parameter dependent systems, likewise, form a set of infinite codimension in the space of vector fields with manifolds of equilibria—if we consider the parameters as fixed phase variables. As classical bifurcation theory is justified by its applicability, so is bifurcation theory without parameters.

This monograph is a slightly extended version of my Habilitation thesis at the Free University Berlin in 2013. I am much indebted to the Free University Berlin in general and to Prof. Dr. Bernold Fiedler in particular for providing an environment where research and teaching, alike, is not only successful but also very enjoyable.

Acknowledgement

This work has been partially supported by the Collaborative Research Center 647 "Space—Time—Matter" of the German Research Foundation (DFG).

Berlin, Germany   Stefan Liebscher
May 2014

Contents

List of Figures

Part I
Preliminaries

Chapter 1
Introduction

This chapter introduces the setting in which we shall study bifurcations without parameters. We compare it with classical bifurcation theory and give an overview and classification of the results presented in the following chapters.

1.1 Classical Bifurcation Versus Bifurcation Without Parameters

We start with a sufficiently smooth vector field. The required smoothness depends on the particular bifurcation problem and will be specified later. We assume a smooth manifold of equilibria, which we can transform to a subspace, at least locally. For simplicity of notation, only, we assume a global flat surface of equilibria although we study a local neighborhood of the origin.

Thus, consider a vector field in an $(n + m)$-dimensional phase space,

$$
\begin{aligned}
\dot{x} &= f(x, y) \in \mathbb{R}^n, \\
\dot{y} &= g(x, y) \in \mathbb{R}^m
\end{aligned}
\tag{1.1}
$$

with an m-dimensional manifold of equilibria $\{ (0, y) \ : \ y \in \mathbb{R}^m \}$, i.e.

$$
f(0, y) \equiv 0, \qquad g(0, y) \equiv 0.
\tag{1.2}
$$

Note the analogy to classical bifurcation theory where y would be a parameter, i.e. $g \equiv 0$:

$$
\begin{aligned}
\dot{x} &= f(x, \lambda) \in \mathbb{R}^n, \qquad f(0, \lambda) \equiv 0, \\
\dot{\lambda} &= 0 \qquad\quad \in \mathbb{R}^m.
\end{aligned}
\tag{1.3}
$$

Here, we write λ instead of y, to emphasize the fact that it is fixed under the flow.

© Springer International Publishing Switzerland 2015
S. Liebscher, *Bifurcation without Parameters*, Lecture Notes in Mathematics 2117,
DOI 10.1007/978-3-319-10777-6_1

Fig. 1.1 A normally
hyperbolic line of equilibria
with flow-invariant foliation

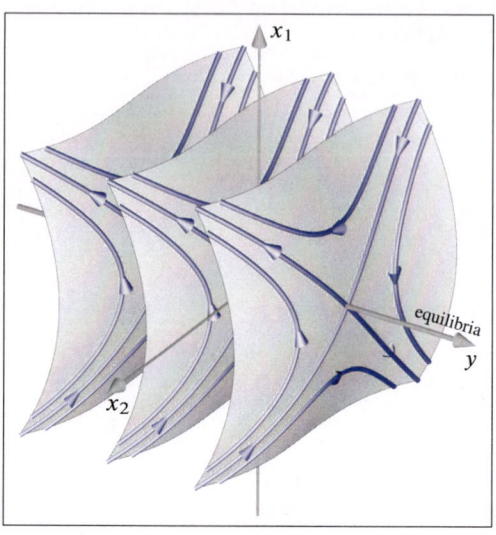

As long as the manifold remains normally hyperbolic, i.e. the linearization of f
in transverse directions on the manifold has no purely imaginary eigenvalues,

$$\text{spec } \partial_x f(0, y) \cap i\mathbb{R} = \emptyset, \tag{1.4}$$

there exists a local flow-invariant foliation with leaves which are homeomorphic
to a standard saddle, for example by the theorem of Shoshitaishvili [65], see also
Fig. 1.1.

Indeed, given an equilibrium of an arbitrary (say differentiable) vector field, [65]
provides a local homeomorphic coordinate transformation such that the transformed
vector field takes the form

$$\begin{aligned}
\dot{x}_+ &= x_+, \\
\dot{x}_c &= f_c(x_c), \\
\dot{x}_- &= -x_-,
\end{aligned} \tag{1.5}$$

where the splitting $x = (x_+, x_c, x_-) \in \mathbb{R}^{n+} \times \mathbb{R}^{n_c} \times \mathbb{R}^{n-}$ corresponds to the splitting
into unstable, neutral and stable eigenspaces of the linearization of the original
vectorfield at the equilibrium. The component f_c in fact corresponds to the reduced
flow (2.2) on the center manifold introduced later. Applied to a normally hyperbolic
equilibrium manifold, the neutral eigenspace is given as the tangent space to the
equilibrium manifold, thus

$$\dot{x}_c = f_c(x_c) \equiv 0, \tag{1.6}$$

yielding a foliation into standard saddles. The local dynamics near a normally
hyperbolic manifold of equilibria is simple, no qualitative changes occur.

Hyperbolic structures are, in general, closely related to structural stability. This observation by Anosov [6] provoked the hope in the 1960s that structural stability could be a generic phenomenon. It also started the whole field of hyperbolic and partially hyperbolic dynamical systems. However, examples of failure of structural stability in open sets of diffeomorphisms due to homoclinic tangencies by Smale [66], also renewed interest in non-generic phenomenons and bifurcations.

Bifurcations are characterized by a breakdown of this normal hyperbolicity. If we write the linearization at the equilibria as

$$\begin{pmatrix} A(y)\ 0 \\ B(y)\ 0 \end{pmatrix} = \begin{pmatrix} \partial_x f(0, y)\ \partial_y f(0, y) \\ \partial_x g(0, y)\ \partial_y g(0, y) \end{pmatrix}, \tag{1.7}$$

then a bifurcation, say at the origin, is characterized by a singular block A, i.e. the spectrum of $A(0)$ intersects the imaginary axis,

$$\mathrm{spec}\,A(0) \cap i\mathbb{R} \neq \emptyset. \tag{1.8}$$

Restricting to a center manifold, see also Sect. 2.1, we can ignore eigenspaces to regular eigenvalues and assume

$$\mathrm{spec}\,A(0) \subset i\mathbb{R}, \tag{1.9}$$

i.e. all eigenvalues at the origin—the bifurcation point—are purely imaginary.

Note the analogy to classical bifurcation theory (1.3). Bifurcations occur at equilibria that have a linearization with eigenvalues on the imaginary axis, i.e. at points $(0, \lambda)$ with $\mathrm{spec}\,\partial_x f(0, \lambda) \cap i\mathbb{R} \neq \emptyset$. For references on classical bifurcation theory see for example [7, 38, 51, 72] and the references there.

In the classical case (1.3), however, the flow-invariant transverse foliation with fibers $\{\lambda = \text{constant}\}$ is also present in a neighborhood of the bifurcation point. In the general case (1.1, 1.2) without parameters, this is no longer true. Indeed, generic functions g of the form (1.2) yield a drift in the "parameter" direction y which excludes any flow-invariant foliation transverse to the manifold of equilibria near a singularity (1.8). Thus, the resulting nonlinear flow profiles differ considerably from classical bifurcation scenarios.

Although we just put large emphasis on the fact that the manifold of equilibria and the bifurcation is not generated by a classical parameter, we are also interested in the mixed cases of m_1-parameter families of m_2-dimensional manifolds of equilibria. These yield foliations due to the m_1 classical parameters. But each fiber still contains an m_2-dimensional manifold of equilibria.

1.2   Manifolds of Equilibria

At a first glance, manifolds of equilibria are a rather degenerate structure. Vector fields with such manifolds form a meager set of infinite codimension in the space of all $\mathscr{C}^k$ vector fields. We will discuss this aspect in Sects. 2.4, 2.5.

At an abstract level we could argue that systems with classical bifurcations are even more degenerate than our setting, due to the additional invariant foliation. Even in this special case, classical bifurcation theory succeeds in discussing many important problems in all areas of dynamical systems.

Bifurcation theory without parameters is necessary to handle many examples in a large variety of applications. Examples include decoupling in networks of coupled oscillators [4, 52], oscillatory viscous in inviscid profiles in hyperbolic balance laws [26, 39], binary oscillations in discretizations [30], population dynamics [24], Bianchi cosmological models [40, 55], stationary profiles in fluid flows [1, 2], memristor dynamics [64], and many more. Chapters 6, 7, 13, 14 are devoted to particular applications.

Although manifolds of equilibria emerge in diverse applications, there is no common physical or mathematical motivation for their appearance. This lack of a common natural cause is another reason for the apparent degeneracy of the setting. Therefore, we take the existence of a manifold of equilibria without transverse flow-invariant foliation as the primary feature. Individual sources of manifolds of equilibria will be discussed together with the exemplary applications.

Of course, there are several structural properties that may generate manifolds of equilibria. However most of them will also induce transverse flow-invariant foliations and represent degenerate cases with respect to our approach. Although they do not fit into the framework discussed here, we briefly introduce important cases to clarify the scope of our setting.

1.2.1   Conserved Quantities

As mentioned before, conserved quantities appear in many applications. Their level sets provide a foliation of the phase space. Equilibria typically form manifolds parametrized by the levels of the conserved quantities, that is, as long as the implicit-function theorem is applicable. At points at which the implicit-function theorem fails, we find bifurcations of the equilibrium set itself.

Bifurcations along manifolds of equilibria generated by conserved quantities are classical bifurcations (the conserved quantities providing classical parameters) and not our aim here.

Note that these conserved quantities can be apparent, for example as the energy function or a continuous symmetry of a Hamiltonian system. Of course, they can also appear as a direct parameter dependence of the model. However in some systems they might be hidden, due to incomplete knowledge of the system and its symmetries, or only exist locally near the bifurcation point.

Such conserved quantities are excluded by non-degeneracy assumptions, or drift-conditions, in all bifurcations analyzed in the following chapters. In fact, these drift conditions will exclude any flow-invariant foliation to lowest possible order of the Taylor expansion of the vector field at the bifurcation point.

1.2.2 Equivariances

Symmetry groups are another structure which is encountered in many models. They are typically given by a Lie group Γ acting on the phase space X and commuting with the flow Φ_t,

$$\Phi_t(\gamma(x)) = \gamma(\Phi_t(x)), \qquad \text{for all } x \in X, \ \gamma \in \Gamma.$$

The corresponding vector field $f = \frac{d}{dt}\Phi_t\big|_{t=0}$ satisfies

$$f(\gamma(x)) = D\gamma(x)f(x), \qquad \text{for all } x \in X, \ \gamma \in \Gamma.$$

Equilibria $x_0 \in X$, $f(x_0) = 0$, come in families given by their group orbits $\Gamma \cdot x_0 = \{\gamma(x_0); \gamma \in \Gamma\}$. For example, for the group $\Gamma = SO(2) = S^1$ of rotations of the plane $X = \mathbb{R}^2$, equilibria form circles around the origin.

In a tubular neighborhood of a group orbit, a Γ-invariant foliation can be constructed. In particular, fibers are parametrized by their intersection points $\gamma(x_0)$ with the group orbit $\Gamma \cdot x_0$. Fibers are also invariant under the stabilizer subgroup $\Gamma_{\gamma(x_0)} = \{\rho \in \Gamma; \rho(\gamma(x_0)) = \gamma(x_0)\}$.

Due to the equivariance of the system, this foliation is also flow invariant. Again, this leads to classical bifurcations, albeit with additional symmetry. See [20, 36] for an introduction into equivariant bifurcation theory.

Whenever we consider additional symmetries in bifurcation problems without parameters, the manifold of equilibria is not an orbit of the symmetry group.

1.2.3 Reversibilities

Time reversibility is another structure that arises frequently. Consider an involution R on $\mathbb{R}^{N+M}$, $R^2 = \text{id}$. For simplicity of notation, we take

$$R(x_1, x_2) = (x_1, -x_2), \qquad x_1 \in \mathbb{R}^{N_1}, x_2 \in \mathbb{R}^{N_2}$$

We call a vector field $f : \mathbb{R}^{N_1+N_2} \to \mathbb{R}^{N_1+N_2}$ time reversible, if for all $x \in \mathbb{R}^{N_1+N_2}$

$$f(Rx) = -Rf(x), \qquad \text{or} \qquad (f_1, f_2)(x_1, -x_2) = (-f_1, f_2)(x_1, x_2).$$

Then the reversibility implies $f_1(x_1, 0) \equiv 0$, and $(x_1, 0)$ is an equilibrium if, and only if, $f_2(x_1, 0) = 0$.

The implicit-function theorem, generically, yields a continuation of the equilibrium $(x_1, 0)$ by a $(N_1 - N_2)$-parameter family of equilibria in the fixed point space of R, for $N_1 > N_2$.

Note however that the linearization at fixed points $x = Rx$ of the reversibility inherits the reversibility

$$Df(x) = -R\, Df(x)\, R.$$

In particular, the spectrum is reflection symmetric to the real and imaginary axes.

Thus, reversibility with a high-dimensional fixed-point space, that is, of dimension higher than half the dimension of the phase space, leads to manifolds of equilibria in the fixed point space of the reversibility. The emerging bifurcations without parameters, however, inherit the reversibility and are not contained in the generic cases discussed in the following chapters. The fully symmetric case of the planar fluid flow, discussed in Chap. 14 and [1], however, is an example of such a reversibility.

1.2.4   Singular Perturbations

Geometric singular perturbation theory is a method to study systems with multiple timescales. In standard form they read

$$\dot{x}_1 = f_1(x_1, x_2),$$
$$\dot{x}_2 = \varepsilon f_2(x_1, x_2),$$

with phase variables $(x_1, x_2) \in \mathbb{R}^{N_1 + N_2}$. The parameter $0 < \varepsilon \ll 1$ separates the two timescales. The formal limit $\varepsilon \to 0$ yields the "fast system"

$$\dot{x}_1 = f_1(x_1, x_2),$$
$$\dot{x}_2 = 0.$$

Its dynamics can be interpreted as fast relaxation to stable sections of the singular manifold $\{x;\ f_1(x) = 0\}$, consisting of equilibria of the fast system.

The fast system, $\varepsilon = 0$, also contains an invariant foliation $\{x_2 = \text{constant}\}$. For $\varepsilon > 0$, on the other hand, f_2 typically induces a slow drift on the singular manifold, modeled by the "slow system", that is the algebro-differential equation

$$0 = f_1(x_1, x_2),$$
$$\dot{x}_2 = f_2(x_1, x_2).$$

Here we expect only isolated equilibria to survive.

The main task of the analysis of singularly perturbed systems is then the combination of solutions of the two formal limit systems to solutions of the full system. Theorems due to Fenichel [25] yield continuations of normally hyperbolic sections of the singular manifold to $\varepsilon > 0$. There, in particular, the foliation of the fast system is transverse to the singular manifold, i.e. bifurcations in the fast system, $\varepsilon = 0$, are classical and the drift for $\varepsilon > 0$ leads to the phenomenon of delayed bifurcation [8].

At tangencies of the fast foliation the singular manifold may break, for $\varepsilon > 0$. Here geometric blow-up or rescaling methods are used to study the full system [48]. Similar methods are employed in our analysis of bifurcations without parameters, see also Sect. 2.6.

1.2.5   Perturbing the Manifold

Without restriction to singularly perturbed problems, the fate of a manifold of equilibria under perturbation of the vector field is often of particular interest. See for example [19] for recent progress on determining emanating branches of equilibria and periodic orbits (as equilibria in suitable function spaces) by Lyapunov-Schmidt reduction.

The aim of this monograph, however, is to study the rich dynamical features near an unperturbed manifold of equilibria—beyond equilibria and periodic orbits. Perturbing the manifold would in particular raise the question of the fate of the intricate structures of heteroclinic orbits that we will find near bifurcations without parameters. Although an interesting aspects of future research, this is beyond the scope of this monograph.

1.2.6   Cosymmetries

Yudovich and Kurakin introduced the concept of cosymmetries [49] to study periodic orbits with Lyapunov-Schmidt reduction in certain PDE problems containing manifolds of equilibria. We discuss the concept in more detail in Chap. 3.

In fact, without additional symmetries or degeneracies, cosymmetries turn out to equivalent to the existence of manifolds of equilibria. Thus the presented theory applies to systems with cosymmetries.

However, the choice of cosymmetries can be ambiguous and the techniques used in this work, in particular center manifolds and normal-form reductions do not preserve particular cosymmetries. Therefore, we do not exploit the cosymmetry structure.

1.3   Classification of Bifurcation Types

Bifurcations without parameters are classified by their codimension; see Sect. 2.5 for a more detailed discussion. The question is: which singularities of the Taylor expansion of a vector field (1.1) can we expect to appear robustly at isolated points along m-dimensional manifolds of equilibria? We call a bifurcation point with such a Taylor expansion *of codimension m*. Our aim is to describe the local dynamics close to the bifurcation.

Analogously, a classical bifurcation of codimension m would appear robustly at isolated parameter values in m-parameter families of vector fields (1.3). We will find the same cases for the transverse linearization $\partial_x f(0, y)$. However, without parameters, the linearization $\partial_y f(0, y)$ might be nonzero, and higher-order Taylor terms typically differ.

In the following, we briefly list the cases of codimension one and two, See Table 1.1 for a complete list including references.

1.3.1   Codimension One

Along one-dimensional manifolds of equilibria, (1.1, $m = 1$), generically at most one algebraically simple eigenvalue zero or a simple pair of nonzero purely imaginary eigenvalues of $\partial_x f(0, y)$ will appear. It crosses the imaginary axis transversely, as y is varied. No singularities of higher order terms arise. We call the arising bifurcations according to their classical counterparts:

- transcritical bifurcation, and
- Poincaré-Andronov-Hopf bifurcation.

Both have been analyzed in earlier papers, [29, 52]. A partial description of the second case can also be found in [24]. For completeness, we discuss them in Chaps. 4 and 5.

1.3.2   Codimension Two

Along two-dimensional manifolds of equilibria, (1.1, $m = 2$), generically the above mentioned bifurcations of codimension one form curves. At isolated points a degeneracy of one higher-order coefficient of the Taylor expansion may appear. We call this

- degenerate transcritical bifurcation, and
- degenerate Poincaré-Andronov-Hopf bifurcation.

Table 1.1 Bifurcations without parameters of codimension one and two

Codimension one, $m = 1$		
$n = 1$: transcritical	$\begin{pmatrix} 0 & 0 \\ \hline 1 & 0 \end{pmatrix}$	Chapter 4 [52]
$n = 2$: Poincaré-Andronov-Hopf	$\begin{pmatrix} 0 & -1 & 0 \\ 1 & 0 & 0 \\ 0 & 0 & 0 \end{pmatrix}$	Chapter 5 [29]
Codimension two, $m = 2$		
$n = 1$: degenerate transcritical	$\begin{pmatrix} 0 & 0 & 0 \\ 0 & 0 & 0 \\ 0 & 0 & 0 \end{pmatrix}$	Chapter 8 [54]
$n = 2$: degenerate Hopf	$\begin{pmatrix} 0 & -1 & 0 & 0 \\ 1 & 0 & 0 & 0 \\ 0 & 0 & 0 & 0 \\ 0 & 0 & 0 & 0 \end{pmatrix}$	Chapter 9
$n = 2$: Bogdanov-Takens	$\begin{pmatrix} 0 & 0 & 0 & 0 \\ 1 & 0 & 0 & 0 \\ 0 & 1 & 0 & 0 \\ 0 & 0 & 0 & 0 \end{pmatrix}$	Chapter 10 [27]
$n = 3$: Zero-Hopf	$\begin{pmatrix} 0 & -1 & 0 & 0 & 0 \\ 1 & 0 & 0 & 0 & 0 \\ 0 & 0 & 0 & 0 & 0 \\ 0 & 0 & 1 & 0 & 0 \\ 0 & 0 & 0 & 0 & 0 \end{pmatrix}$	Chapter 11
$n = 4$: Hopf-Hopf	$\begin{pmatrix} 0 & -1 & 0 & 0 & 0 & 0 \\ 1 & 0 & 0 & 0 & 0 & 0 \\ 0 & 0 & 0 & -\omega & 0 & 0 \\ 0 & 0 & \omega & 0 & 0 & 0 \\ 0 & 0 & 0 & 0 & 0 & 0 \\ 0 & 0 & 0 & 0 & 0 & 0 \end{pmatrix}$	Chapter 12

The table lists the dimension m of the manifold of equilibria of (1.1), the cross-sectional dimension n, the normal form of the linearization (1.7), and references

These bifurcations are described in Chaps. 8 and 9. Note that although the names of the bifurcation types are chosen in analogy to classical bifurcations, their normal forms differ and so do their degeneracies. We will also provide a brief comparison to the classical bifurcations.

Alternatively, the linearization $\partial_x f(0, y)$ can be of codimension two. It can possess either an algebraically double and geometrically simple eigenvalue zero, an algebraically simple eigenvalue together with a simple pair of purely imaginary eigenvalues, or two non-resonant simple pairs of purely imaginary eigenvalues. We find

- Bogdanov-Takens bifurcation,
- Zero-Hopf bifurcation, and
- Hopf-Hopf bifurcation.

The Bogdanov-Takens bifurcation without parameters has been analyzed earlier, [27], and is described in Chap. 10. The other two bifurcations are discussed in Chaps. 11 and 12, although results are less complete than on the other cases and leave room for further investigation and development.

We study codimension-two bifurcations not only in the "pure" setting of a generic two-dimensional manifold of equilibria but also in the "mixed" setting of a one-parameter family of one-dimensional manifolds of equilibria. The latter case also yields a two-dimensional manifold of equilibria in the extended phase space, but endowed with flow-invariant foliation. Each fiber contains a one-dimensional manifolds of equilibria. The mixed cases can be seen as a bridge between the classical bifurcations with isolated equilibria in the fibers and the pure bifurcations without parameters with no fibers at all. In fact, we find mixed and pure cases to be quite similar for Bogdanov-Takens and Zero-Hopf bifurcations, but very different for degenerate transcritical, degenerate Poincaré-Andronov-Hopf, and Hopf-Hopf bifurcations.

1.4   Further Cases

Bifurcations of codimension three and higher are still open for research. One exception is the case of bifurcations of arbitrary codimension along manifolds of equilibria with only one cross-sectional direction, i.e. m-dimensional manifolds of equilibria in $(m + 1)$-dimensional phase space. These bifurcation correspond to singularity theory of vector fields on the real line, see [54] and Chap. 15.

Applications often display additional structure. For example, symmetries of the original problem can give rise to equivariances of the dynamical system and change the bifurcation pictures. An example is discussed in Chap. 14, a reversible Bogdanov-Takens bifurcation without parameters.

Another example is the correspondence of a rotationally symmetric Poincaré-Andronov-Hopf bifurcation to a transcritical bifurcation with additional reflection symmetry, also called pitchfork bifurcation. Even without rotational symmetry of the original problem, the truncated normal form of a Poincaré-Andronov-Hopf bifurcation yields this symmetry, see also Sect. 2.2 and Chap. 5.

Chapter 2
Methods and Concepts

In this chapter we present basic concepts and methods used in the analysis of bifurcations.

2.1 Center Manifolds

Center manifolds facilitate the reduction of the dimension of a bifurcation problem to the necessary minimum. The local center manifold of an equilibrium, i.e. the bifurcation point, is a smooth manifold tangential to the center eigenspace of that equilibrium. The center eigenspace is the generalized eigenspace to all purely imaginary (or zero) eigenvalues of the linearization of the vector field at the equilibrium. The local center manifold contains all bounded solution in a small neighborhood, in particular all equilibria, all periodic orbits, and all connecting (heteroclinic) orbits of equilibria. It contains all the features that characterize the local flow near a bifurcation point. Trajectories outside the center manifold follow a corresponding trajectory on the center manifold with a saddle-type dynamics in the cross-sectional directions. This is also called a *slaving principle*.

Theorem 2.1 *Consider a $\mathscr{C}^k$ vector field*

$$\dot{x} = f(x) = Ax + \tilde{f}(x) \qquad \in \mathbb{R}^n$$

with equilibrium at the origin, $f(0) = 0$. Let $A = Df(0)$ be the linearization at the origin and $\tilde{f} = \mathscr{O}(\|x\|^2)$ the nonlinear terms.

Let $\mathbb{R}^n = E^u \oplus E^s \oplus E^c$ be the eigenspace decomposition with respect to unstable, stable and critical eigenvalues of A, i.e. $E^{u/s/c}$ are invariant under A and all eigenvalues of A restricted to $E^{u/s/c}$ have positive/negative/zero real parts.

Then there exist a local C^k manifold W^c, tangential to E^c at $x = 0$, of the same dimension and locally invariant i.e. everywhere tangential to the vector field.

© Springer International Publishing Switzerland 2015
S. Liebscher, *Bifurcation without Parameters*, Lecture Notes in Mathematics 2117,
DOI 10.1007/978-3-319-10777-6_2

Furthermore, W^c contains all solutions that stay in a small enough neighborhood of the origin for all times $t \in \mathbb{R}$.

Proofs can be found in [41, 72]. The idea of the proof is to switch to a global statement for a nonlinearity with sufficiently small $\mathscr{C}^1$ norm, by a suitable cut-off function. Then the variation-of-constants formula, projected onto stable, unstable and center component by $\Pi^{u/s/c}$,

$$
\begin{aligned}
x^u(t) &= \int_t^\infty e^{A(t-s)} \Pi^u \tilde{f}(x(s)) \ ds \\
x^s(t) &= \int_{-\infty}^t e^{A(t-s)} \Pi^s \tilde{f}(x(s)) \ ds \\
x^c(t) &= x_0^c + \int_0^t e^{A(t-s)} \Pi^c \tilde{f}(x(s)) \ ds
\end{aligned}
\tag{2.1}
$$

provides a contraction mapping on the space of functions $x(\cdot)$ with exponentially weighted norm, the weight chosen between zero and the smallest absolute value of nonzero real parts of eigenvalues of A. The fixed point $x^*(x_0; \cdot)$ then provides the center manifold $W^c = \{ x^*(x_0^c; 0)) \mid x_0^c \in E^c \}$.

For the abstract analysis of bifurcations, this theorem justifies assumption (1.9) that all eigenvalues of the linearization at the bifurcation point lie on the imaginary axis.

In applications this constitutes the first step of the analysis: the reduction of the problem to the center manifold. In fact, the calculation of the center manifold $x^{u,s} = x^{u,s}(x^c)$ and the reduced vector field $f^{red} : E^c \to E^c$ can be done simultaneously using the invariance of the manifold and its tangency to the eigenspace,

$$
f(x^{u,s}(x^c), x^c) = \begin{pmatrix} Dx^{u,s}(x^c) \\ \mathrm{id} \end{pmatrix} f^{red}(x^c).
\tag{2.2}
$$

Note that the reduced vector field still contains the manifold of equilibria which we started with. The reduced vector field has arbitrary but finite smoothness, bounded by the smoothness of the original vector field. An additional smooth coordinate transformation, bounded by the smoothness of the manifold, straightens the manifold of equilibria. We arrive at the setting (1.1, 1.2, 1.9) of the introduction.

2.2 Normal Forms

Analysis of the local dynamics near an equilibrium exploits the Taylor-expansion of the vector field at the equilibrium. This expansion, however, depends on the chosen coordinate system. The first step is therefore the choice of *good coordinates*.

But what are good coordinates? One possible answer is: good are coordinates which yield the simplest possible Taylor expansion of the vector filed: we want as many coefficients of the Taylor expansion as possible to vanish. This is the usual point of view of normal-form theory. One general normal-form algorithm is described below.

Unfortunately, the simplest possible Taylor expansion is usually not suited best for later analysis. Firstly, given additional structures should be respected. Here, this is mainly the manifold of equilibria. A modified normal-form algorithm is discussed in the next section. Secondly, hidden structures often become visible only in modified coordinates at the expense of a higher number of nonzero Taylor coefficients. For example, a Hamilton structure to leading order greatly facilitates the analysis of the Bogdanov-Takens bifurcation in Chap. 10.

We use normal forms as presented in [72]. See also [62]. The basic idea is to eliminate terms of the Taylor expansion of a vector field $F(z) = Az + F_2(z) + F_3(z) + \cdots$ by a coordinate transformation $z = \Psi(\tilde{z}) = \tilde{z} + \Psi_2(\tilde{z}) + \Psi_3(\tilde{z}) + \cdots$, given as its Taylor series. Here, F_k, Ψ_k denote homogeneous polynomials of degree k. We find the transformed vector field $\tilde{F}$ as

$$D\Psi(\tilde{z})\tilde{F}(\tilde{z}) = F(\Psi(\tilde{z})). \tag{2.3}$$

Taylor terms of order k yield

$$\tilde{F}_k(\tilde{z}) = F_k(\tilde{z}) + A\Psi_k(\tilde{z}) - D\Psi_k(\tilde{z})A\tilde{z} + R(\tilde{z}) \tag{2.4}$$

where the remainder R contains only terms in F_ℓ, $\tilde{F}_\ell$, Ψ_ℓ with $2 \leq \ell < k$.

We can therefore successively eliminate components of $F_k(z)$ in the range of ad A,

$$((\text{ad } A)\Psi_k)(z) = [A, \Psi_k](z) = A\Psi_k(z) - D\Psi_k(z)Az. \tag{2.5}$$

The normal form of F is then given, up to any finite order k, by a linear complement to the range of ad A. Note that the elimination step k will create additional terms of higher orders.

The correct choice of complements depends on the problem. However, with a suitable scalar product in the space of homogeneous vector polynomials, the choice

$$\ker(\text{ad } A)^{\mathrm{T}} = \ker \text{ad}\,(A^{\mathrm{T}}) \tag{2.6}$$

yields a complement which is easy to calculate. Although it might not be tuned to the problem, this choice of complement has an additional benefit: the normal form terms $\tilde{G}_k$, $k \geq 2$, commute with the group generated by A^{T},

$$e^{A^{\mathrm{T}}t}G_k(z) = G_k(e^{A^{\mathrm{T}}t}z). \tag{2.7}$$

If the linearization A is normal, $AA^\mathrm{T} = A^\mathrm{T}A$, so is the normal form $Az + G_2(z) + \ldots + G_k(z)$. This additional normal-form symmetry, alone, might greatly facilitate the analysis of a problem. The most prominent example is the rotational symmetry of the normal form of Poincaré-Andronov-Hopf points due to the pair of imaginary eigenvalues, see Chap. 9.

2.3 Normal Forms with Manifolds of Equilibria

In our setting, we start with a manifold of equilibria. Unfortunately, this manifold is not preserved by the normal-form algorithm presented in the previous section. We can restrict the coordinate transformations used in the normal-form algorithm to those that fix the manifold. The resulting normal form then has more non-vanishing coefficients but retains the straight manifold of equilibria. Unfortunately, the remaining nonzero coefficients do not depend solely on the linearization but also on the manifold of equilibria, i.e. on the particular bifurcation problem.

For $z = (x, y)$ with equilibrium set $\{x = 0\}$ we must restrict our coordinate transformations $\Psi = (\Psi^x, \Psi^y)$ to those with

$$\Psi^x(0, y) = 0. \tag{2.8}$$

Then, the transformed vector field will retain the set $\{x = 0\}$ of equilibria.

In [27] this adjusted normal-form procedure is carried out in detail for the Bogdanov-Takes bifurcation without parameters, see also Chap. 10. Strictly speaking, however, it is not necessary for the analysis there. The rescaling procedure which is used following the normal-form reduction would yield the same result with a much cruder initial simplification of the vector field, see Chap. 10.

2.4 Genericity

Genericity is the topological notion of large sets. We call a subset $U \subset X$ of a complete metric space *generic*, if it is the intersection of countably many open and dense sets,

$$U = \bigcap_{k \in \mathbb{N}} U_k, \qquad \mathrm{int}\, U_k = U_k, \qquad \mathrm{clos}\, U_k = X.$$

Due to the Baire category theorem, a generic set is still dense. Countable intersections of generic sets are still generic.

Complements of generic sets are called *meager*. A meager set is the union on countably many nowhere dense sets, or of closed sets without interior,

$$V = \bigcup_{k \in \mathbb{N}} V_k, \qquad \text{int clos } V_k = \emptyset.$$

A countable union of meager sets can never cover the whole (complete metric) space.

A generic vector field thus means an arbitrary vector field from a generic subset of the space of all vector fields, typically of a given smoothness. The generic subset is usually specified by several non-degeneracy conditions, forcing certain coefficients of the Taylor expansion of the vector field to be nonzero. In this case the specified generic set is even open and dense.

For example, a generic linear map $A : \mathbb{R}^n \to \mathbb{R}^n$ is hyperbolic, i.e. has no eigenvalues on the imaginary axis. In other words: generically, no bifurcations occur. In fact, the set of linear maps $A : \mathbb{R}^n \to \mathbb{R}^n$ without purely imaginary eigenvalues is open and dense in the space of all linear maps $\mathbb{R}^n \to \mathbb{R}^n$.

Genericity heavily depends on the considered space. In fact, the next section will introduce the framework to find generic bifurcations.

2.5 Unfoldings and Codimension

For smooth one-parameter families of linear maps, families of linear maps $A(s) : \mathbb{R}^n \to \mathbb{R}^n$ without eigenvalues on the imaginary axis are not generic any more. Indeed, take $n = 1$, $A(0) = 0$, $A'(0) = 1$, then by the implicit-function theorem, every family $\tilde{A}(\cdot)$ in a sufficiently small neighborhood of $A(\cdot)$ has a zero.

The codimension of a singularity is defined as follows. A given linear map $A_0 : \mathbb{R}^n \to \mathbb{R}^n$ has codimension m, if there is a generic subset U of the set of all smooth m-parameter families

$$\{ A : (-\varepsilon, \varepsilon)^m \to L(\mathbb{R}^n, \mathbb{R}^n) \mid A(0) = A_0 \},$$

such that every family $A(\cdot) \in U$ has a neighborhood V such that for every family $\tilde{A}(\cdot) \in V$ there exists s_0 and an invertible linear map Φ with

$$A_0 = \Phi^{-1} \tilde{A}(0) \Phi.$$

In other words: every sufficiently small perturbation $\tilde{A}(\cdot)$ of a generic m-parameter unfolding $A(\cdot)$ of A_0 has an element $\tilde{A}(s_0)$ which is equivalent to A_0.

In our bifurcation analysis A will be the linearization at the bifurcation restricted to the center eigenspace. Genericity of U will be phrased in transversality and non-degeneracy conditions of the already provided family $A(\cdot)$ of linearizations along the manifold of equilibria or by additional parameters.

Furthermore, we will not only discuss singularities of the linear part, but also of terms of higher order in the Taylor expansion. Then we require that we can recover all the singular Taylor terms after perturbation. Equivalence will be mediated by a nonlinear coordinate transformation or normal-form reduction.

This approach to singularities, their unfoldings, their codimension, and their classification is the starting point of singularity theory or catastrophe theory, see also [8].

2.6   Rescaling and Blow Up

A successful method to study the local dynamics of a vector field

$$\dot{x} = f(x), \qquad x \in \mathbb{R}^n, \qquad f : \mathbb{R}^n \to \mathbb{R}^n, \tag{2.9}$$

near a singularity $x = 0$ is to rescale the vector field to blow up and thereby desingularize the singularity. The method is also called quasi-homogeneous rescaling.

This is achieved by a vector $\alpha \in \mathbb{N}_+^n$ of positive integers and the transformation

$$x = \sigma^\alpha(\tilde{x}) := \operatorname{diag}(\sigma^\alpha)\, \tilde{x} = (\sigma^{\alpha_1}\tilde{x}_1, \ldots, \sigma^{\alpha_n}\tilde{x}_n), \tag{2.10}$$

for small $0 < \sigma \ll 1$. Every sufficiently small neighborhood of the origin in the old coordinates is also contained in a small neighborhood of the origin in the new coordinates. In particular, all small bounded trajectories of (2.9) are also small bounded trajectories of

$$\dot{\tilde{x}}_k = \sigma^{-\alpha_k} f_k(\sigma^\alpha(\tilde{x})), \qquad k = 1, \ldots, n. \tag{2.11}$$

Let $\alpha_* \in \mathbb{N}$ be the leading order, that is the minimal exponent of σ among all monomials with nonzero coefficients in the Taylor expansion of the vector field (2.11). Typically, $\alpha_* > 0$, if the origin is a singularity of the original vector field. Then, the rescaling of time $t = \sigma^{-\alpha_*}\tilde{t}$ yields the system

$$\tilde{x}_k' = \sigma^{-\alpha_k - \alpha_*} f_k(\sigma^\alpha(\tilde{x})), \qquad k = 1, \ldots, n. \tag{2.12}$$

In particular, the Taylor expansion of (2.11) with respect to σ starts with terms of order 0 in σ.

Summarized: the coordinates are rescaled by (positive) powers of the rescaling parameter σ. Then the resulting vector field is divided by σ to the largest possible power, without introducing a singularity in σ.

The limiting system (2.12, $\sigma = 0$) corresponds to a desingularized vector field of the blown-up singularity $x = 0$. Regular perturbation theory can be applied to obtain results for $\sigma \gtrsim 0$, describing the dynamics in a neighborhood of $x = 0$.

Good choices for the scaling α are given by the Newton polyhedron. Let

$$f_k = \sum_{\beta \in \mathbb{N}^n} c_{k,\beta} x^{\beta} = \sum_{\beta \in \mathbb{N}^n} c_{k,\beta} x_1^{\beta_1} \cdots x_n^{\beta_n}, \qquad k = 1, \ldots, n, \qquad (2.13)$$

be the Taylor expansion of the vector field. Then the Newton polyhedron is the convex hull of the powers of monomials with nonzero coefficients of the vector field

$$N := \operatorname{conv} \{ \beta - e_k \mid c_{k,\beta} \neq 0; \ k = 1, \ldots, n; \ \beta \in \mathbb{N}^n \}, \qquad (2.14)$$

where $e_k = (0, \ldots, 0, 1, 0, \ldots, 0)$ is the k-th unit vector. The adjustment by e_k accounts for the factor $\sigma^{-\alpha_k}$ in (2.11).

Every outer facet F of N, facing the origin, yields a viable scaling

$$\alpha \perp F. \qquad (2.15)$$

The time rescaling α_* is then given by the distance of F from the origin. In fact, when such α, α_* are used, the leading order system (2.12, $\sigma = 0$) contains exactly the monomials to points of N in F.

An alternative point of view is to consider (2.12) for $0 < \sigma \ll 1$ and $\|\tilde{x}\| = 1$ as spherical coordinates near the origin. Then the boundary $\sigma = 0$ is the blow up of the singularity $x = 0$ to a sphere $\|\tilde{x}\| = 1$. The boundary vector field (2.12, $\sigma = 0$) on this sphere is expected to be less singular then the original vector field at the origin [23]. Recursive blow ups can (and have been) used to further desingularize the vector field. This technique has been successfully used not only to study local trajectories but also to get quantitative results on the passage of trajectories close to singularities [46].

Chapter 3
Cosymmetries

Cosymmetries have been introduced by Yudovich and Kurakin to study limit cycles near manifolds of equilibria via Lyapunov-Schmidt reduction [49, 50]. They turn out to be equivalent to the existence of manifolds of equilibria, provided some non-degeneracy conditions are satisfied.

Given a vector field $F : \mathbb{R}^n \to \mathbb{R}^n$, a cosymmetry is any other vector field orthogonal to F.

$$L : \mathbb{R}^n \to \mathbb{R}^n; \qquad \text{such that for all } x \in \mathbb{R}^n \quad \langle F(x), L(x) \rangle = 0. \qquad (3.1)$$

A non-cosymmetric equilibrium is any zero of F where the cosymmetry does not vanish.

$$F(x_0) = 0, \qquad L(x_0) \neq 0. \qquad (3.2)$$

Then, necessarily, the adjoint of the linearization of the vector field has the non-trivial kernel vector $L(x_0)$,

$$0 = \langle DF(x_0)\xi, L(x_0) \rangle + \langle F(x_0), DL(x_0)\xi \rangle = \langle DF(x_0)\xi, L(x_0) \rangle, \qquad (3.3)$$

thus the linearization $DF(x_0)$ has a nontrivial kernel, too.

Theorem 3.1 *Let the origin be a non-cosymmetric equilibrium of a $\mathscr{C}^k$ vector field F, $k \geq 1$. Let the kernel of the linearization be one-dimensional, Then the set of equilibria of F near the origin forms a one-parameter $\mathscr{C}^k$ curve.*

Proof Let φ, ψ be unit kernel vectors of the linearization and its adjoint,

$$\ker DF(0) = \operatorname{span}\{\varphi\}, \quad \|\varphi\| = 1, \qquad \psi = L(0)/\|L(0)\|.$$

© Springer International Publishing Switzerland 2015
S. Liebscher, *Bifurcation without Parameters*, Lecture Notes in Mathematics 2117,
DOI 10.1007/978-3-319-10777-6_3

Step 1:   We claim that equilibria of F, close to the origin, are given by zeros of the orthogonal projection onto the complement of ψ, i.e.

$$0 \;=\; F(x) \qquad \Longleftrightarrow \qquad 0 \;=\; \Pi_{\psi^\perp} F(x) \;=\; F(x) - \langle F(x), \psi \rangle \psi.$$

One direction is trivial. Therefore assume $0 = \Pi_{\psi^\perp} F(x)$. Then $F(x) = \alpha \psi$ with coefficient $\alpha \in \mathbb{R}$. In particular $\langle \alpha L(0), L(x) \rangle = 0$. Continuity of L, $L(0) \neq 0$ implies $\alpha = 0$ for x close to the origin.

Step 2:   Consider the map $\tilde{F} = \Pi_{\psi^\perp} F : \mathbb{R}^n \to L(0)^\perp \cong \mathbb{R}^{n-1}$. Then $D\tilde{F}(0)$ has full rank, $\operatorname{rank} D\tilde{F}(0) = \operatorname{rank} DF(0) = n - 1$. Thus, the implicit-function theorem yields the claim. $\square$

We also see that the curve of equilibria is given by the curve $x(s) = s\varphi + \tilde{x}(s)$ with unique $\tilde{x} \perp \varphi$.

Theorem 3.2 *Consider a $\mathscr{C}^k$ vector field $F : \mathbb{R}^n \to \mathbb{R}^n$ with a line of equilibria, w.l.o.g. $F(x_1, 0, \ldots, 0) \equiv 0$. Assume that the kernel of the linearization at the origin is one-dimensional. Then, locally near the origin, there exists a cosymmetry $L : \mathbb{R}^n \to \mathbb{R}^n$ of F, such that the origin is a non-cosymmetric equilibrium.*

Proof Let φ, ψ be unit kernel vectors of the linearization and its adjoint,

$$\ker DF(0) = \operatorname{span}\{\varphi\}, \quad \varphi = (1, 0, \ldots, 0), \qquad \operatorname{image} DF(0) \perp \psi, \quad \|\psi\| = 1.$$

Due to the x_1 axis of equilibria we can decompose $F(x) = K(x)x$, where the matrix $K(x)$ has a first column of zeros. Thus, for arbitrary L we have the equivalence

$$0 = \langle F(x), L(x) \rangle \qquad \Longleftrightarrow \qquad 0 = \langle x, K^*(x) L(x) \rangle$$

with adjoint K^*. Note that K is non-unique but image $K^*(x) = \phi^\perp \cong \mathbb{R}^{n-1}$.

To construct a cosymmetry $L(x) = \psi + \tilde{L}(x)$, we need to solve $0 = K^*(x)(\psi + \tilde{L})$, i.e. we look for zeros of the map

$$T : \mathbb{R}^n \times \psi^\perp \to \mathbb{R}^{n-1}, \qquad (x, \tilde{L}) \; \to \; K^*(x)(\psi + \tilde{L}).$$

Note that $\psi^\perp \cong \mathbb{R}^{n-1}$.

We find $T(0, 0) = 0$, and $D_{\tilde{L}} T(0, 0) = K^*(0)$ of full rank. Again, the implicit-function theorem yields the claim. In particular, the constructed cosymmetry has normalized projection onto the kernel of the adjoint of the linearization. $\square$

Both theorems can be extended to sets of m simultaneous cosymmetries and m-dimensional manifolds of equilibria. A non-cosymmetric equilibrium for cosymmetries $L_1, \ldots, L_m$ is then a point x_0, such that

$$F(x_0) \;=\; 0, \qquad \dim \operatorname{span}\{ L_k(x_0) \,|\, k = 1, \ldots, m \} \;=\; m. \tag{3.4}$$

The non-degeneracy condition on the vector field reads $\dim \ker DF(x_0) = m$.

The condition on the kernel of the linearizations is in fact consistent with the non-degeneracy conditions of our bifurcations discussed in the following chapters, as long as no additional symmetries are considered. Additional symmetries might enlarge the kernel as in Sect. 4.2, such that the above theorems do not apply.

Unfortunately, we are not able to exploit the additional structure of cosymmetries in our bifurcation analysis as it conflicts with our normal form and rescaling approaches.

Part II
Codimension One

Chapter 4
Transcritical Bifurcation

In this chapter we study the simplest bifurcation without parameters: a line of equilibria which loses normal stability when a simple eigenvalue crosses zero transversely. This case has already been studied in [52], see also [28].

Note that we assume the system to already be reduced to the center manifold, see assumption (1.9) and the center-manifold theorem in Sect. 2.1. All results are local and aim at a description of the dynamics in a small enough neighborhood of the bifurcation point.

4.1 The Generic Case

In classical bifurcation theory, a transcritical bifurcation of a primary equilibrium arises in one-parameter families. In a two-dimensional center manifold

$$\dot{x} = f(x,\lambda) \in \mathbb{R}, \qquad f(0,\lambda) \equiv 0,$$
$$\dot{\lambda} = 0 \qquad \in \mathbb{R} \tag{4.1}$$

an eigenvalue zero of the linearization, say at the origin,

$$0 = \partial_x f(0,0) \tag{4.2}$$

generically crosses zero transversely as λ increases: $0 \neq \partial_\lambda \partial_x f(0,0)$. Without loss of generality, we take

$$\partial_\lambda \partial_x f(0,0) > 0. \tag{4.3}$$

© Springer International Publishing Switzerland 2015
S. Liebscher, *Bifurcation without Parameters*, Lecture Notes in Mathematics 2117,
DOI 10.1007/978-3-319-10777-6_4

a   **b**

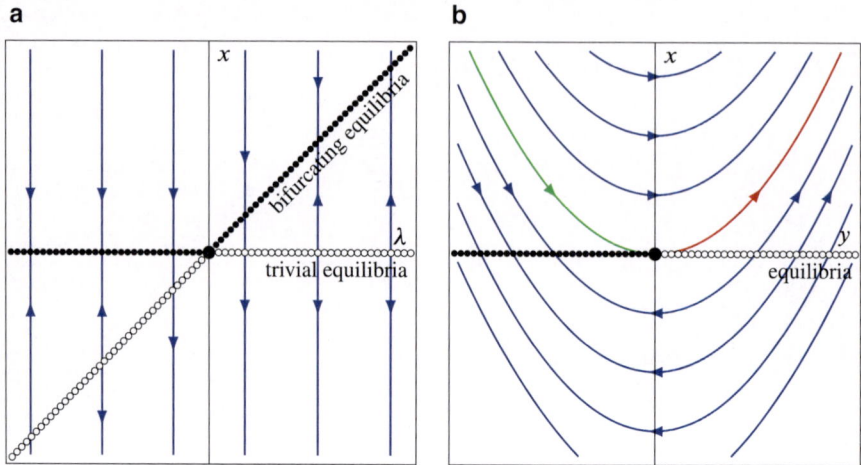

Fig. 4.1 Transcritical bifurcation. (**a**) Classical bifurcation with bifurcating branch of equilibria. (**b**) Without parameters, stable manifold of the origin in *green*, unstable manifold in *red*

Assuming the additional non-degeneracy condition

$$\partial_x^2 f(0,0) \neq 0, \tag{4.4}$$

there exists a $\mathscr{C}^2$ coordinate change $\tilde{\lambda} = \tilde{\lambda}(\lambda)$, $\tilde{x} = \tilde{x}(x, \lambda)$, transforming system (4.1) to the normal form (tildes omitted)

$$\dot{x} = x(\lambda - x). \tag{4.5}$$

See Fig. 4.1a.

Without parameters,

$$\begin{aligned}
\dot{x} &= f(x, y) & \in \mathbb{R}, && f(0, y) \equiv 0, \\
\dot{y} &= g(x, y) & \in \mathbb{R}, && g(0, y) \equiv 0,
\end{aligned} \tag{4.6}$$

the nontrivial eigenvalue $\partial_x f(0, y)$ can change sign along the line of equilibria $\{y = 0\}$,

$$\partial_x f(0,0) = 0. \tag{4.7}$$

Generically, it will do so transversely,

$$\partial_y \partial_x f(0,0) > 0. \tag{4.8}$$

The non-degeneracy condition, however, is replaced with

$$\partial_x g(0,0) \neq 0 \tag{4.9}$$

and yields a two-dimensional Jordan block of the linearization at the transcritical point. Indeed, as y is not a parameter in our setting, generically it is subject to a drift to lowest possible order in the Taylor expansion.

Theorem 4.1 (Transcritical Bifurcation Without Parameters) *[29] Consider a $\mathscr{C}^2$ vector field with a curve of equilibria. Assume the curve loses normal stability due to a real eigenvalue zero (4.7). Assume the generic transversality and non-degeneracy conditions (4.8, 4.9) in a two-dimensional center manifold.*

Then, (on the center manifold) there exists a (local) $\mathscr{C}^1$-diffeomorphism which (locally) maps orbits of the vector field (4.6) to orbits of the normal form

$$\begin{aligned} \dot{x} &= xy, \\ \dot{y} &= x, \end{aligned} \tag{4.10}$$

with preserved time orientation.

In a local neighborhood of the transcritical bifurcation point, trajectories form parabolas tangent to the line of equilibria at the transcritical point. The flow direction is reversed on opposite sides of the equilibrium line. See Fig. 4.1b.

Proof The vector field vanishes identically on the equilibrium manifold, $f(0, y) \equiv 0$, $g(0, y) \equiv 0$, see (4.6), and we have only one transverse direction, $x \in \mathbb{R}$. This allows us to factor out x,

$$\begin{aligned} \dot{x} &= f(x, y) = x\tilde{f}(x, y), \\ \dot{y} &= g(x, y) = x\tilde{g}(x, y). \end{aligned} \tag{4.11}$$

with $\mathscr{C}^1$-functions $\tilde{f}, \tilde{g}$. This system has the same orbits as the rescaled system

$$\begin{aligned} x' &= \tilde{f}(x, y), \\ y' &= \tilde{g}(x, y). \end{aligned} \tag{4.12}$$

except for the line of equilibria at $x = 0$ and for the reversed flow direction for $x < 0$. Conditions (4.7, 4.9) translate to

$$\tilde{f}(0,0) = 0, \qquad \partial_y \tilde{f}(0,0) > 0, \qquad \tilde{g}(0,0) \neq 0. \tag{4.13}$$

By the flow-box theorem, we can locally transform (4.12) to

$$\begin{aligned} \tilde{x}' &= 0, \\ \tilde{y}' &= 1. \end{aligned} \tag{4.14}$$

Due to (4.13), the y-axis is transformed to the curve

$$\tilde{x} = p(\tilde{y}) = a\tilde{y}^2 + \text{h.o.t.}, \tag{4.15}$$

with $a \neq 0$. Indeed, let Φ_t be the flow to (4.12), then a suitable transformation is given by $(x, y) = h(\tilde{x}, \tilde{y}) := \Phi_{\tilde{y}}(\tilde{x}, 0)$. We have $\partial_x h(0, 0) = (1, 0)$, thus the implicit-function theorem yields the solution curve $(0, x) = h(p(\tilde{y}), \tilde{y})$ with $p(0) = 0$, $p'(0) = -\tilde{f}(0, 0) = 0$, and $2a = p''(0) = -\partial_y \tilde{f}(0, 0) \cdot \tilde{g}(0, 0) \neq 0$.

Finally, the $\mathscr{C}^1$-diffeomorphism

$$
\begin{aligned}
\hat{x} &= -\text{sign}(a)\tilde{x}, \\
\hat{y} &= \tilde{y}\sqrt{|p(\tilde{y})|\tilde{y}^{-2}} = \sqrt{|a|}\tilde{y} + \text{h.o.t.}
\end{aligned}
\tag{4.16}
$$

preserves the flow lines $\{\tilde{y} = \text{constant}\}$ of (4.14) and transforms the curve (4.15) to the parabola $\hat{x} = -\hat{y}^2$. The coordinate change (4.16) applied after the flow-box transformation yields the claimed normal form (4.10). This proves the theorem. $\square$

4.2   Additional Reflection Symmetry

In this section we investigate the loss of normal stability of a line of equilibria by a simple eigenvalue zero as in the last section. However, we assume an additional reflection symmetry of the system. This provides the simplest example of an equivariant bifurcation without parameters and prepares the discussion of the Poincaré-Andronov-Hopf bifurcation in the next chapter.

In classical bifurcation theory, this corresponds to a pitchfork bifurcation of a primary equilibrium with one parameter. In a two-dimensional center manifold

$$
\begin{aligned}
\dot{x} &= f(x, \lambda) \quad &\in \mathbb{R}, \qquad f(0, \lambda) \equiv 0, \\
\dot{\lambda} &= 0 \quad &\in \mathbb{R}
\end{aligned}
\tag{4.17}
$$

we assume an additional equivariance with respect to a reflection,

$$f(-x, \lambda) = -f(x, \lambda), \qquad \text{for all } x, \lambda. \tag{4.18}$$

Again, an eigenvalue crosses zero transversely at the origin,

$$0 = \partial_x f(0, 0), \qquad 0 < \partial_\lambda \partial_x f(0, 0). \tag{4.19}$$

The equivariance (4.18) forces the value of $\partial_x^2 f(0, 0)$ to vanish, therefore the former non-degeneracy condition (4.4) is replaced with

$$0 \neq \partial_x^3 f(0, 0). \tag{4.20}$$

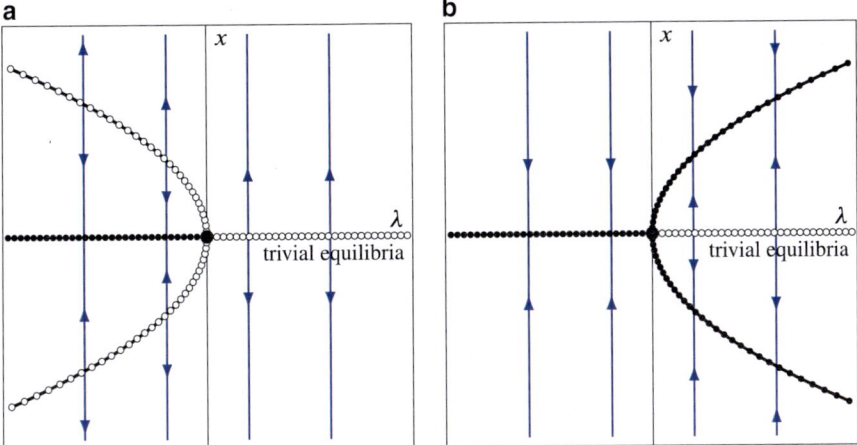

Fig. 4.2 Classical pitchfork bifurcation. (**a**) Subcritical case with bifurcating branch of unstable equilibria. (**b**) Supercritical case with bifurcating branch of stable equilibria

The resulting normal form reads

$$\dot{x} = x(\lambda \pm x^2). \tag{4.21}$$

Depending on the sign of $\partial_x^3 f(0,0)\partial_\lambda\partial_x f(0,0)$ the pitchfork bifurcation is called subcritical (positive sign) or supercritical (negative sign). See Fig. 4.2.

Without parameters, the system reads

$$\begin{aligned} \dot{x} &= f(x,y) \in \mathbb{R}, & f(0,y) &\equiv 0, \\ \dot{y} &= g(x,y) \in \mathbb{R}, & g(0,y) &\equiv 0, \end{aligned} \tag{4.22}$$

with the same equivariance with respect to a reflection in x,

$$\left.\begin{aligned} f(-x,y) &= -f(x,y) \\ g(-x,y) &= g(x,y) \end{aligned}\right\} \quad \text{for all } x, y. \tag{4.23}$$

Again, we assume a transverse eigenvalue crossing,

$$0 = \partial_x f(0,0), \qquad 0 < \partial_y\partial_x f(0,0). \tag{4.24}$$

The new non-degeneracy condition,

$$0 \neq \partial_x^2 g(0,0), \tag{4.25}$$

again generates a drift along the line of equilibria to lowest possible order. Note that the linearization $\partial_x g(0,0)$ vanishes due to the equivariance (4.23) of the system.

Theorem 4.2 ($\mathbb{Z}_2$-Equivariant Transcritical Bifurcation) *[29] Consider a $\mathscr{C}^2$ vector field with a curve of equilibria. Assume the curve loses normal stability due to a real eigenvalue zero. Let the vector field be equivariant with respect to a reflection which leaves the curve of equilibria pointwise fixed. Assume the generic transversality and non-degeneracy conditions (4.24, 4.25) in a two-dimensional center manifold with nontrivial action of the equivariance.*

Then, in (the center manifold) there exists a (local) $\mathscr{C}^1$-diffeomorphism which (locally) maps orbits of the vector field (4.22, 4.23) to orbits of the normal form

$$
\begin{aligned}
\dot{x} &= xy, \\
\dot{y} &= \tfrac{1}{2}\delta x^2, \qquad \delta = \pm 1,
\end{aligned}
\tag{4.26}
$$

with preserved time orientation.

We call $\delta = +1$ the hyperbolic and $\delta = -1$ the elliptic case. In the hyperbolic case, there are no small bounded solutions close to the bifurcation point, except the given curve of equilibria. In the elliptic case a local neighborhood of the transcritical bifurcation point in the center manifold is filled with heteroclinic connections from the unstable part to the stable part of the given curve of equilibria. See Fig. 4.3.

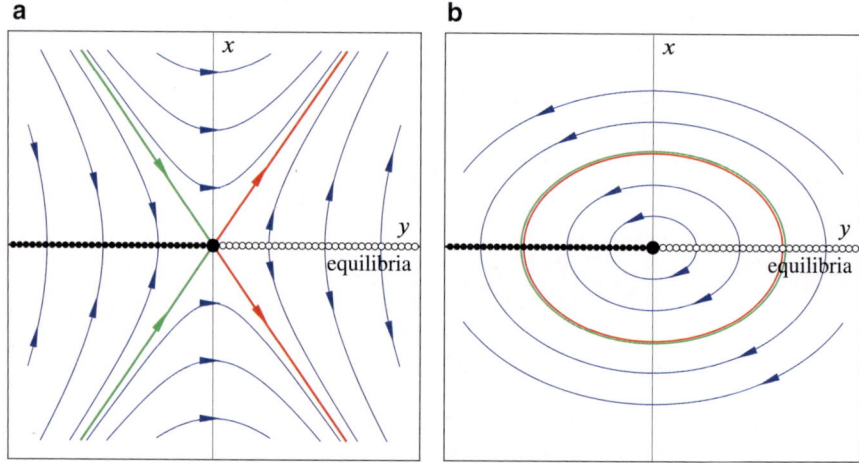

Fig. 4.3 Transcritical bifurcation with reflection symmetry. (**a**) Hyperbolic case, stable manifold of the origin in *green*, unstable manifold in *red*. (**b**) Elliptic case, stable manifold of an exemplary left equilibrium, in *green*, and unstable manifold of corresponding right equilibrium, in *red*, coincide identically

Proof We start similar to the proof of theorem (4.10): in (4.22) we factor out x and obtain

$$\dot{x} = f(x, y) = x\tilde{f}(x, y),$$
$$\dot{y} = g(x, y) = x\tilde{g}(x, y),$$

(4.27)

with $\mathscr{C}^1$-functions $\tilde{f}, \tilde{g}$. This system has the same orbits as the rescaled system

$$x' = \tilde{f}(x, y),$$
$$y' = \tilde{g}(x, y).$$

(4.28)

except for the line of equilibria at $x = 0$ and for the reversed flow direction for $x < 0$. Conditions (4.24, 4.25) translate to

$$\tilde{f}(0, 0) = 0, \qquad \partial_y \tilde{f}(0, 0) > 0, \qquad \partial_x \tilde{g}(0, 0) \neq 0,$$

(4.29)

whereas the symmetry assumption (4.23) yields a time reversibility

$$\tilde{f}(-x, y) = \tilde{f}(x, y)$$
$$\tilde{g}(-x, y) = -\tilde{g}(x, y)$$

(4.30)

with respect to the involution $R(x, y) = (-x, y)$. In particular, the involution R maps solutions of (4.28) onto solutions with reversed time.

Note that $\tilde{g}(0, y) \equiv 0$ due to reversibility (4.30). In particular $(0, 0)$ is an equilibrium since $\tilde{f}(0, 0) = 0$ due to (4.29), and $\partial_y \tilde{g}(0, 0) = 0$. Thus we can set

$$\delta = -\text{sign det } D \begin{pmatrix} f \\ g \end{pmatrix} (0, 0) = \text{sign } \partial_y \tilde{f}(0, 0) \partial_x \tilde{g}(0, 0) = \pm 1.$$

(4.31)

A simple rescaling of x, y then yields the normalized linearization

$$D \begin{pmatrix} f \\ g \end{pmatrix} (0, 0) = \begin{pmatrix} 0 & 1 \\ \delta/2 & 0 \end{pmatrix}.$$

(4.32)

In the hyperbolic case, $\delta = +1$, the origin is a hyperbolic equilibrium of (4.28). This not only justifies the name but also yields a $\mathscr{C}^1$-coordinate transformation

$$\tilde{x} = \tilde{x}(x, y) = x + \cdots, \qquad \tilde{y} = \tilde{y}(x, y) = y + \cdots,$$

(4.33)

that linearizes the vector field, due to Belitskii's theorem [12]. The averaged transformation

$$\hat{x} = (\tilde{x}(x, y) - \tilde{x}(-x, y))/2, \qquad \hat{y} = (\tilde{y}(x, y) + \tilde{y}(-x, y))/2,$$

(4.34)

commutes with the reversibility R and still linearizes the vector field, due to the reversibility of the vector field. Furthermore, the averaged transformations leaves the fixed-point space of R, i.e. line of equilibria, fixed, $\hat{x}(0, y) = 0$. Hence it provides the claimed orbit equivalence between (4.22) and the normal form (4.26), in the hyperbolic case.

In the elliptic case, $\delta = +1$, the origin is a center equilibrium of (4.28). Reversible Hopf bifurcation [72] yields a local family $\gamma_s(t) \in \mathbb{R}^2$ of periodic orbits surrounding the origin and parametrized over $s > 0$, such that $\gamma_s(0) = (s, 0)$ on the y-axis. Alternatively, we can construct this family directly: through C^1 dependence on parameters, every orbit close to the origin has to follow the linearized flow, i.e. the harmonic oscillator, for finite time, and thus hits the y-axis at least twice. Every orbit intersecting the fixed-point space of the reversibility R, on the other hand, is a reversible periodic orbit. In particular it is mapped by R onto itself.

By C^1 dependence on initial values, the passage time from fix(R) to fix(R) and thereby the minimal period of γ_s is given by a C^1 function $p(s) > 0$. We have chosen $\gamma_s(0)$ to lie on the positive y-axis, i.e. the fixed-point space of R. By reversibility again, $\gamma_s(p(s)/2)$ must also lie in this fixed-point space. The orbit spends half a period above and half a period below the y-axis, both parts being images of each other under R. Closeness to the linearized flow forces $\gamma_s(p(s)/2)$ to the negative y-axis.

The transformation

$$s \begin{pmatrix} \sin 2t \\ 2 \cos 2t \end{pmatrix} \longmapsto \gamma_s(p(s)t/\pi) \tag{4.35}$$

now maps the orbits of the linearized flow onto the orbits of (4.28) and maps the y-axis onto itself. Hence it provides the claimed orbit equivalence between (4.22) and the normal form (4.26), in the elliptic case.   □

Chapter 5
Poincaré-Andronov-Hopf Bifurcation

In classical bifurcation theory, a Poincaré-Andronov-Hopf bifurcation arises in one-parameter families of real vector fields, when a pair of conjugate complex eigenvalues crosses the imaginary axis, as the parameter varies. A family of periodic orbits will bifurcate. To be specific, in a three-dimensional center manifold

$$\begin{aligned} \dot{x} &= f(x, \lambda) &\in \mathbb{R}^2, \qquad f(0, \lambda) \equiv 0, \\ \dot{\lambda} &= 0 &\in \mathbb{R} \end{aligned} \tag{5.1}$$

a purely imaginary pair of eigenvalues of the linearization, say at the origin,

$$\begin{pmatrix} 0 & 1 \\ -1 & 0 \end{pmatrix} = \partial_x f(0, 0) \tag{5.2}$$

generically crosses the imaginary axis transversely as λ increases $0 \neq \partial_\lambda \operatorname{div}_x f(0, 0)$. Without loss of generality, we take

$$0 < \partial_\lambda \operatorname{div}_x f(0, 0) = \partial_\lambda (\partial_{x_1} f_1(0, 0) + \partial_{x_2} f_2(0, 0)). \tag{5.3}$$

A truncated normal form is then given by

$$\dot{z} = (\lambda + i + c|z|^2)z \tag{5.4}$$

in complex notation, $z \in \mathbb{C}$ with complex coefficient $c \in \mathbb{C}$. This normal form is also called Stuart-Landau oscillator. Assuming non-degenerate real part of c,

$$c_{\Re e} := \Re e \, c \neq 0, \tag{5.5}$$

© Springer International Publishing Switzerland 2015
S. Liebscher, *Bifurcation without Parameters*, Lecture Notes in Mathematics 2117,
DOI 10.1007/978-3-319-10777-6_5

we find a family of stable periodic orbit around the unstable equilibrium for $\lambda > 0$ in the supercritical case, $\mathfrak{Re}\, c < 0$, or a family of unstable periodic orbit around the stable equilibrium for $\lambda < 0$ in the subcritical case, $\mathfrak{Re}\, c > 0$.

In polar coordinates $z = r e^{i\varphi}$, we could also write

$$
\begin{aligned}
r' &= (\lambda + c_{\mathfrak{Re}}\, r^2) r \\
\varphi' &= 1 + c_{\mathfrak{Im}}\, r^2.
\end{aligned}
\tag{5.6}
$$

Ignoring the φ-component, close to constant rotation, we find a classical pitchfork bifurcation in the radius, see Sect. 4.2 and Fig. 4.2. Due to the rotation in φ, the bifurcating equilibria of the pitchfork represent bifurcating periodic orbits. Further details on classical Poincaré-Andronov-Hopf bifurcation can be found in [59, 72]. Without parameters, Poincaré-Andronov-Hopf points have been studied in [29]:

Theorem 5.1 ([29]) *Let* $F : \mathbb{R}^N \to \mathbb{R}^N$ *be a* $\mathscr{C}^5$ *vector field with a line of fixed points along the* u_1-axis, $F(u_1, 0, \ldots, 0) \equiv 0$. *At* $u_1 = 0$, *we assume the Jacobi matrix* $DF(u_1, 0, \ldots, 0)$ *to be hyperbolic, except for a trivial kernel vector along the* u_1-axis *and a* complex *conjugate pair of simple, purely imaginary, nonzero eigenvalues* $\mu(u_1), \overline{\mu(u_1)}$ *crossing the imaginary axis transversely as* u_1 *increases through* $u_1 = 0$:

$$
\begin{aligned}
\mu(0) &= i\omega(0), \qquad \omega(0) > 0, \\
\mathfrak{Re}\, \mu'(0) &\neq 0.
\end{aligned}
\tag{5.7}
$$

Let Z *be the two-dimensional real eigenspace of* $F'(0)$ *associated to* $\pm i\omega(0)$. *By* Δ_Z *we denote the Laplacian with respect to variations of* u *in the eigenspace* Z. *Coordinates in* Z *are chosen as coefficients of the real and imaginary parts of the complex eigenvector associated to* $i\omega(0)$. *Note. that the linearization acts as a rotation with respect to these coordinates, which are not necessarily orthogonal. Let* P_0 *be the one-dimensional eigenprojection onto the trivial kernel along the* u_1-axis. *Our final non-degeneracy assumption then reads*

$$
\Delta_Z P_0 F(0) \neq 0.
\tag{5.8}
$$

Fixing orientation along the positive u_0-axis, *we can consider* $\Delta_Z P_0 F(0)$ *as a real number. Depending on the sign*

$$
\eta := \mathrm{sign}\,(\mathfrak{Re}\, \mu'(0)) \cdot \mathrm{sign}\,(\Delta_Z P_0 F(0)),
\tag{5.9}
$$

we call the Hopf point $u = 0$ *elliptic if* $\eta = -1$ *and hyperbolic for* $\eta = +1$.

Then the following holds true in a neighborhood U *of* $u = 0$ *within a three-dimensional center manifold to* $u = 0$.

In the hyperbolic case, $\eta = +1$, *all non-equilibrium trajectories leave the neighborhood* U *in positive or negative time direction (possibly both). The stable and unstable sets of* $u = 0$, *respectively, form cones around the positive/negative*

a

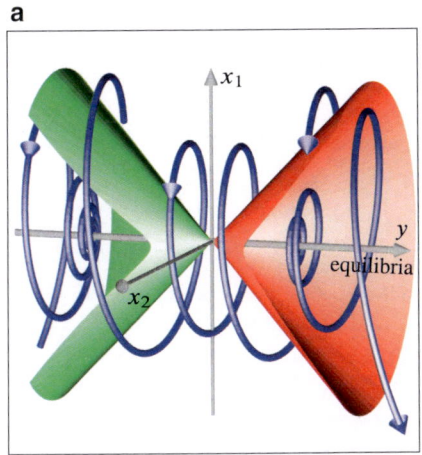

b

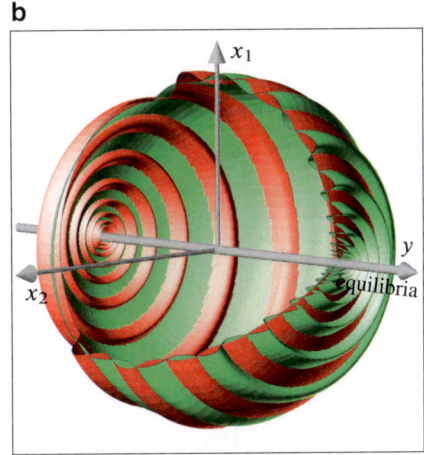

Fig. 5.1 Poincaré-Andronov-Hopf bifurcation without parameters. (**a**) Hyperbolic case, stable manifold of the origin in green, unstable manifold in *red*. (**b**) Elliptic case, stable manifold of an exemplary left equilibrium in *green*, unstable manifold of an exemplary right equilibrium in *red*; the manifolds are cut open for better visibility

u_1-*axis, with asymptotically elliptic cross section near their tips at $u = 0$. These cones separate regions with different convergence behavior. See Fig. 5.1a.*

In the elliptic case all non-equilibrium trajectories starting in U are heteroclinic between equilibria $u^\pm = (u_1^\pm, 0, \ldots, 0)$ on opposite sides of the Hopf point $u = 0$. If $F(u)$ is real analytic near $u = 0$, then the two-dimensional strong stable and strong unstable manifolds of $u^\pm$ within the center manifold intersect at an angle which possesses an exponentially small upper bound in terms of $|u^\pm|$. See Fig. 5.1b.

The formulation of the assumptions of the above theorem can be simplified: we first restrict to the three dimensional center manifold and assume that this manifold is flat. Then we take coordinates in direction of the real, generalized eigenvectors of the linearization at the Hopf point. To this end, consider a system

$$\begin{pmatrix} \dot{x} \\ \dot{y} \end{pmatrix} = F(x, y) = \begin{pmatrix} f(x, y) \\ g(x, y) \end{pmatrix}, \qquad x \in \mathbb{R}^2, \quad y \in \mathbb{R}, \qquad (5.10)$$

$x = (x_1, x_2)$, $f = (f_1, f_2)$, with the following properties:

(i) There exists a line of equilibria, $F(0, y) \equiv 0$.
(ii) The origin has pair of purely imaginary, nonzero eigenvalues in transverse direction to the equilibrium plane:

$$\partial_x f(0, 0) = \begin{pmatrix} 0 & -1 \\ 1 & 0 \end{pmatrix}.$$

(iii) This nontrivial eigenvalue pair crosses the imaginary axis with nonvanishing speed as y increases, $\partial_y \mathrm{div}_x f(0,0) > 0$.
(iv) There is a drift along the line of equilibria, that is g satisfies the following non-degeneracy condition:

$$\eta = \mathrm{sign}\, \Delta_x g(0,0) \neq 0,$$

with $\Delta_x = \partial_{x_1}^2 + \partial_{x_2}^2$.

The first condition is our structural assumption, (ii) describes our bifurcation point, and (iii,iv) are non-degeneracy assumptions fulfilled generically. Note the correspondence of (iii,iv) to (5.7, 5.9). We normalized the purely imaginary eigenvalue to $\pm i$ by rescaling time. We normalized $\partial_x f(0,0)$ by choosing the real generalized eigenvectors as a basis in x. The sign of (iii) is fixed by reflecting y, if necessary.

This setup is robust, i.e. under small perturbations of F respecting (i) there is a point near the origin satisfying (ii–iv) for the perturbed system. From the point of view of singularity theory, this is indeed a singularity of codimension one, which is unfolded versally by the coordinate y along the line of trivial equilibria.

Note how the bifurcation is determined by quadratic terms alone: the transversality (ii) and the drift (iv). The coefficient of the cubic term that distinguishes subcritical from supercritical classical Hopf bifurcation, see (5.5), is replaced by the quadratic drift-term η distinguishing elliptic from hyperbolic Hopf points without parameters.

The proof of Theorem 5.1 can then be sketched as follows.

The normal-form procedure, see Sects. 2.2, 2.3, and [72], yields a normal form which is equivariant w.r.t. rotations $\{\exp(DF(0,0)^\mathsf{T}\tau)\ ;\ \tau \in \mathbb{R}\}$ up to an arbitrary but finite order of the Taylor expansion. We obtain the normal form in polar coordinates $(x_1 + i x_2) = r \exp(i\varphi)$:

$$
\begin{aligned}
\dot{r} &= ry + rh^r(y, re^{i\varphi}), \\
\dot{\varphi} &= 1, \\
\dot{y} &= \eta r^2 + rh^y(y, re^{i\varphi}).
\end{aligned}
\tag{5.11}
$$

The error terms h^r, h^y are of order $\mathcal{O}((|y| + |r|)^K)$ beyond the (arbitrary) normal form order K. The factor r in front (h^r, h^y) accounts for the vanishing of these error terms along the line of equilibria.

Ignoring φ and the error terms for the moment, this is the $\mathbb{Z}_2$-equivariant transcritical bifurcation discussed in Sect. 4.2.

We "only" have to superimpose the rotation in φ. Thus we re-interpret the flow profiles of Fig. 4.3 as pictures of the Poincaré return map to a fixed cross section. Orbits limiting at equilibria then represent stable/unstable manifold of these equilibria.

The rigorous discussion of higher-order terms (h^r, h^y) not in normal form and thereby breaking the rotational normal-form symmetry constitutes the main part of

[29]. To this end we introduce spherical polar coordinates $(y + ir) = R \exp(i\vartheta)$ to blow up the bifurcation point. The normal form then reads

$$
\begin{aligned}
\dot{R} &= (R \sin \vartheta)^2 \cos \vartheta \cdot (1 + \eta) + R^2 \sin \vartheta \cdot h^R(R, \vartheta, \varphi), \\
\dot{\vartheta} &= (R \sin \vartheta)(\cos^2 \vartheta - \eta \sin^2 \vartheta) + R \sin \vartheta \cdot h^\vartheta(R, \vartheta, \varphi), \\
\dot{\varphi} &= 1,
\end{aligned}
\tag{5.12}
$$

with error terms (h^R, h^ϑ) of order $\mathcal{O}(R^{K-1})$ for $R \searrow 0$.

We want to rescale time by R to blow up the bifurcation point at the origin. This however induces a rapid oscillation of the angle φ, close to the line of equilibria at $R = 0$. Therefore we rescale both, for $R > 0$:

$$
\begin{aligned}
\psi &= R^2 \varphi, \\
d\tau &= R \, dt.
\end{aligned}
\tag{5.13}
$$

The angle ψ is considered in $\mathbb{R}$ as the universal cover of S^1. Denoting $' = d/d\tau$ and abbreviating $c := \cos \vartheta$, $s := \sin \vartheta$, the normal form (5.12) now reads

$$
\begin{aligned}
R' &= Rs(cs(1 + \eta) + h^R), \\
\vartheta' &= s(c^2 - \eta s^2 + h^\vartheta), \\
\psi' &= R + 2\psi s(cs(1 + \eta) + h^R).
\end{aligned}
\tag{5.14}
$$

Note that the error terms $(h^R, h^\vartheta) = (h^R, h^\vartheta)(R, \vartheta, R^{-2}\psi)$ now rapidly oscillate in ψ but are still of order $\mathcal{O}(R^{K-1})$ for $R \searrow 0$.

The domain $\Omega = \{R > 0, \ 0 < \vartheta < \pi\}$ is invariant. The closure of $\Omega \cap \{R < \varepsilon\}$, for small $\varepsilon > 0$, corresponds to a full neighborhood of the Hopf point in the original system (5.10). In particular, R cannot become zero in finite time. Thus, in this domain, $t \to \pm\infty$ is equivalent to $\tau \to \pm\infty$. The Hopf point corresponds to the boundary $\{R = 0\}$, the line of equilibria to the boundary $\{s = 0\}$. The intersection $\{R = 0, s = 0\}$ still consists of equilibria. It suffices to study system (5.14) in $\Omega \cap \{R < \varepsilon\}$ to prove Theorem 5.1.

In the elliptic case, $\eta = -1$, we find $R' = Rsh^R$ to be small and $\vartheta' = s(1 + h^\vartheta)$ of constant sign. This yields a transition from ϑ near 0 to ϑ near π, in finite time τ, with arbitrarily little drift in R due to the error terms. In fact, R'/R is also small compared to ϑ', therefore $\lim_{\tau \to -\infty} = 0$, $\lim_{\tau \to +\infty} = \pi$ while the relative drift in R remains arbitrarily small. This proves the claimed heteroclinic connections in the elliptic case. We will discuss splitting of separatrices at the end of this section.

In the hyperbolic case, $\eta = +1$, we find additional equilibria at $(R, \vartheta, \psi) = (0, \vartheta_\pm^*, 0)$ with

$$
\cos^2 \vartheta_\pm^* = \sin^2 \vartheta_\pm^*, \qquad \text{that is} \quad \sin \vartheta_\pm^* = \tfrac{1}{2}\sqrt{2}.
$$

The angles $\vartheta_{\pm}^{*}$ relate to the asymptotic angles of the conical stable/unstable sets of the bifurcation point, in original variables. For normal-form order $K \geq 4$ the linearization at the additional equilibria is given by

$$\begin{pmatrix} \pm\frac{1}{2}\sqrt{2} & 0 & 0 \\ * & \mp\sqrt{2} & 0 \\ 1 & 0 & \pm\sqrt{2} \end{pmatrix}. \tag{5.15}$$

In particular, these equilibria are strictly hyperbolic with associated stable and unstable manifolds. The saddle point property causes solutions sufficiently close to these equilibria to converge in forward (or backward) time, or else get ejected along the unstable (or stable) manifolds.

The two-dimensional unstable manifold W_{+}^{u} of $(0, \vartheta_{+}^{*}, 0)$ constitutes the unstable set of the Hopf point. The two-dimensional stable manifold W_{-}^{s} of $(0, \vartheta_{-}^{*}, 0)$ constitutes the stable set of the Hopf point. The one-dimensional stable/unstable manifolds lie in the boundary $R = 0$—the blown-up Hopf point—and are inaccessible for trajectories in Ω. Now consider again any trajectory in the domain $\Omega \cap \{R < \varepsilon\}$, not contained in the above manifolds. Then the trajectory must remain in one of the three regions of $\Omega \cap \{R < \varepsilon\}$ bounded by W_{+}^{u} and W_{-}^{s}.

In forward time, a trajectory cannot get arbitrarily close to ϑ_{-}^{*}, as it would be ejected along the one-dimensional unstable manifold in ϑ-direction. Therefore, in fact, $|\vartheta - \vartheta_{-}^{*}|$ remains bounded from below, away from zero. A trajectory getting close to ϑ_{+}^{*} is ejected along W_{+}^{u}: the value of R becomes large and the trajectory leaves the neighborhood of the Hopf point. In all other cases, the argument of the elliptic case applies: the trajectory converges to $\{s = 0\}$, i.e. it connects to the line of equilibria.

In backward time, the analogous argument yields ejection along W_{-}^{s} or convergence to the line of equilibria.

Let us return to the elliptic case. We found heteroclinic connections as the truncated normal form ((5.11), $h = 0$) and Fig. 4.3 suggest. However, beware of the identically coinciding stable/unstable manifolds near the elliptic Hopf points. Without φ-dependence, we have a flow in the plane, and intersections must contain a trajectory. With φ-dependence, we have a Poincaré map on the plane, and intersections are generically transverse. For the full system near an elliptic Hopf point, separatrices of the Poincaré map split as sketched in Fig. 5.2 and the three-dimensional views of Fig. 5.1.

Neishtadt's theorem on exponential averaging [63] provides upper bounds on the size of the splitting of strong stable/unstable manifolds. This upper bound is exponentially small in the distance from the origin, for analytic vector fields F. Lower bounds on the splitting are not established. See also [31] on further results on the splitting of separatrices in close-to-integrable systems and [33] on the difficult question of lower bounds on the splitting of separatrices.

Neishtadt's averaging method uses the complex extension of the analytic vector field to obtain bounds on its derivatives. This provides bounds on the error terms

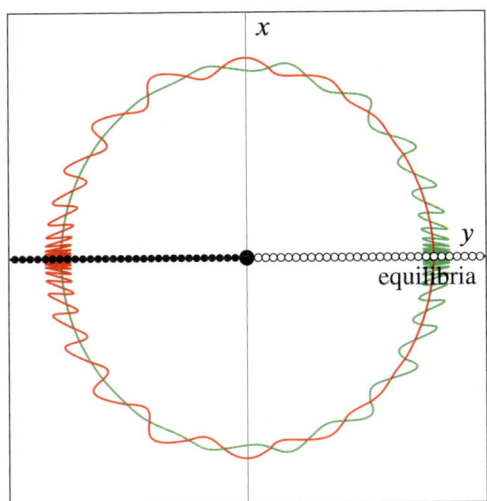

Fig. 5.2 Poincaré-Andronov-Hopf bifurcation, splitting of separatrices. Elliptic case, $\eta < 0$. Compare with Fig. 4.3b. Stable manifold of a selected left equilibrium in *green*, unstable manifold of a selected right equilibrium in *red*. The separatrices split due to higher-order terms of the Poincaré return map that break the rotational equivariance. Stable/unstable manifolds of all other equilibria look similar

of the normal form (5.11) in an ε-neighborhood of the bifurcation point. In fact, by careful balancing of error terms, this can be done up to normal-form order $N = \mathcal{O}(1/\varepsilon)$. The truncated normal form (independent of the angle φ) then differs from the full system only by an exponentially small term.

Applied to the Hopf bifurcation without parameters, exponential upper bounds on the splitting are established in [29]. Near the Hopf point we expect heteroclinic connections of any unstable equilibrium to an entire interval of stable equilibria on the other side of the Hopf point. For analytic vector fields, the size of the interval is exponentially small in the distance from the Hopf point, by Neishtadt averaging. Lower bounds on the size of the interval are, however, unknown.

Chapter 6
Application: Decoupling in Networks

Networks are an important structure in many applications ranging from chemistry and biology to engineering. Pattern formation in networks has caught an ever growing interest in recent years [57]. The main focus is usually the synchronization of the cells of the network. Here, we study the converse phenomenon: under suitable symmetry assumptions, networks can decouple and continua of states emerge where all couplings cancel out each other and several pairs of cells can have arbitrary phase differences.

Let us consider a square, an octahedron, or a general graph Γ of $2m$ vertices $\{\pm 1, \ldots, \pm m\}$, such that each vertex k is connected with every other vertex except the antipode $-k$, see Fig. 6.1. Let this graph represent the additive couplings between cells,

$$\dot{u}_k = f_k\left(u_k, \sum_{\ell \neq \pm k} u_\ell\right). \tag{6.1}$$

Think of individual cells as single oscillators: without coupling, they move along a limit cycle. This will be made precise below.

Assume a group of equivariances generated by the exchange of individual antipodal pairs together with a sign switch:

$$f_{-k}(-u_k, 0) = -f_k(u_k, 0), \qquad 1 \le k \le m. \tag{6.2}$$

Due to the additive coupling, the effects of antipodal cells on their neighbors cancel, if the antipodal cells have opposite values. Due to the symmetry, antipodal cells remain opposite if the total coupling vanishes. Therefore, the antipode space

$$\Sigma := \{ u = (u_k)_{1 \le \pm k \le m} \mid u_{-\ell} = -u_\ell \text{ for all } \ell \} \tag{6.3}$$

© Springer International Publishing Switzerland 2015
S. Liebscher, *Bifurcation without Parameters*, Lecture Notes in Mathematics 2117,
DOI 10.1007/978-3-319-10777-6_6

Fig. 6.1 Network of coupled
oscillators

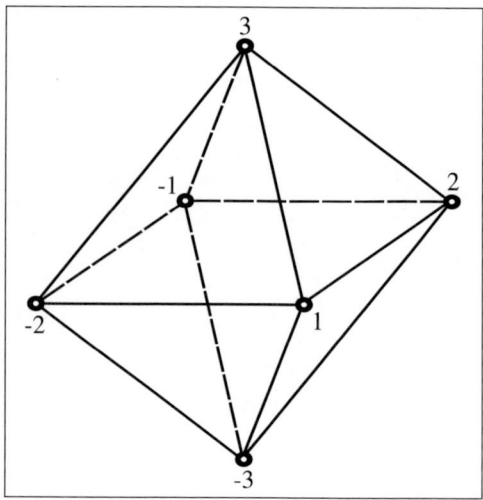

is invariant under the flow of (6.1). In fact, on Σ all cells u_k decouple and we find
the direct product flow of m antipodal pairs $u_k = -u_{-k}$

$$\dot{u}_k = f_k(u_k, 0), \qquad 1 \le k \le m. \tag{6.4}$$

This decoupling phenomenon was first observed in [4, 5].

Assume further, that every individual cell dynamics (6.4) possesses a periodic
orbit $\gamma_k(t)$ with a common fixed period 2π. The simplest example would be
provided by identical cells, $f_k = f$, in the plane, with an additional S^1-
equivariance. In complex notation, $u_k \in \mathbb{C}$, the equivariance is given by:

$$f(e^{i\varphi} u) = e^{i\varphi} f(u), \qquad \text{for all } \varphi \in \mathbb{R}. \tag{6.5}$$

Then we can choose arbitrary phase angles φ_k, $k = 1, \ldots, m$ to find an m-
dimensional torus of periodic solutions

$$\dot{u}_k(t) = \gamma_k(t + \varphi_k), \qquad 1 \le k \le m. \tag{6.6}$$

in the decoupling subspace Σ. These periodic solutions yield an $(m-1)$-dimensional
manifold of fixed points of the Poincaré return map to a section $\{t + \varphi_m = 0$
$(\text{mod } 2\pi)\}$. With S^1 equivariance (6.5) the flow (6.1) can be pulled back by the
symmetry group to a flow on the Poincaré section, and the set (6.6) of periodic orbits
becomes an $(m-1)$-dimensional manifold of equilibria of this pulled back flow.

These equilibria describe states of the network where the couplings add up to zero
at each cell, i.e. the cells do not "see" each other anymore. Different antipodal pairs
can have arbitrary phase differences. Nevertheless, as it turns out below, although
the coupling vanishes at the decoupled states, the coupling can still stabilize them.

In [52] this program has been carried out for the square ring of four identical S^1-equivariant oscillators with linear coupling:

$$\dot{z}_k = (f(|z_k|) + i)z_k + \alpha e^{i\chi}(z_{k-1} + z_{k+1}) \quad \in \mathbb{C}, \quad k = 1,\dots,4 \ (\text{mod } 4),$$
(6.7)

with $f \in \mathbb{R}$, $f(1) = 0$, $f'(1) = -1$, $\alpha > 0$. In polar coordinates $z_k = r_k e^{i\phi_k}$, this yields

$$\begin{aligned}
\dot{r}_k &= f(r_k) + \alpha r_{k+1} \cos(\phi_{k+1} - \phi_k + \chi) + \alpha r_{k-1} \cos(\phi_{k-1} - \phi_k + \chi), \\
\dot{\phi}_k &= g \quad + \alpha \tfrac{r_{k+1}}{r_k} \sin(\phi_{k+1} - \phi_k + \chi) + \alpha \tfrac{r_{k-1}}{r_k} \sin(\phi_{k-1} - \phi_k + \chi),
\end{aligned}$$
(6.8)

and the decoupling subspace is given by

$$D = \{(r_1,\dots,r_4,\phi_1,\dots,\phi_4) \mid r_k = r_{k+2}, \ \phi_i = \phi_{k+2} + \pi \ (\text{mod } 2\pi), \ i = 1,\dots,4\}.$$

On D, all coupling effects add up to zero and the pairs of antipodal cells decouple, i.e. on D the system has product structure:

$$\begin{aligned}
\dot{r}_1 &= f(r_1,0), \ \dot{\phi}_1 = g; \ r_3 = r_1, \ \phi_3 = \phi_1 + \pi \ (\text{mod } 2\pi); \\
\dot{r}_2 &= f(r_2,0), \ \dot{\phi}_2 = g; \ r_4 = r_2, \ \phi_4 = \phi_2 + \pi \ (\text{mod } 2\pi).
\end{aligned}$$
(6.9)

To factor out the S^1-symmetry, we replace the angles by their differences

$$\begin{aligned}
\psi_1 &= \tfrac{1}{2}(-\phi_1 + \phi_3) - \tfrac{1}{2}\pi, \\
\psi_2 &= \tfrac{1}{2}(+\phi_2 - \phi_4) - \tfrac{1}{2}\pi, \\
\rho &= \tfrac{1}{2}(-\phi_1 + \phi_2 - \phi_3 + \phi_4) - \tfrac{1}{2}\pi,
\end{aligned}$$
(6.10)

and find

$$\begin{aligned}
\dot{r}_1 &= f(r_1) + \alpha r_2 \sin(+\psi_1 + \psi_2 + \rho + \chi) - \alpha r_4 \sin(+\psi_1 - \psi_2 + \rho + \chi), \\
\dot{r}_2 &= f(r_2) + \alpha r_3 \sin(+\psi_1 - \psi_2 - \rho + \chi) - \alpha r_1 \sin(-\psi_1 - \psi_2 - \rho + \chi), \\
\dot{r}_3 &= f(r_3) + \alpha r_4 \sin(-\psi_1 - \psi_2 + \rho + \chi) - \alpha r_2 \sin(-\psi_1 + \psi_2 + \rho + \chi), \\
\dot{r}_4 &= f(r_4) + \alpha r_1 \sin(-\psi_1 + \psi_2 - \rho + \chi) - \alpha r_3 \sin(+\psi_1 + \psi_2 - \rho + \chi),
\end{aligned}$$

$$\begin{aligned}
\dot{\psi}_1 &= +\frac{\alpha}{2}\frac{r_2}{r_1}\cos(+\psi_1 + \psi_2 + \rho + \chi) - \frac{\alpha}{2}\frac{r_4}{r_1}\cos(+\psi_1 - \psi_2 + \rho + \chi) \\
&\quad - \frac{\alpha}{2}\frac{r_4}{r_3}\cos(-\psi_1 - \psi_2 + \rho + \chi) + \frac{\alpha}{2}\frac{r_2}{r_3}\cos(-\psi_1 + \psi_2 + \rho + \chi),
\end{aligned}$$

$$\dot{\psi}_2 = -\frac{\alpha}{2}\frac{r_3}{r_2}\cos(+\psi_1 - \psi_2 - \rho + \chi) + \frac{\alpha}{2}\frac{r_1}{r_2}\cos(-\psi_1 - \psi_2 - \rho + \chi)$$

$$+ \frac{\alpha}{2} \frac{r_1}{r_4} \cos(-\psi_1 + \psi_2 - \rho + \chi) - \frac{\alpha}{2} \frac{r_3}{r_4} \cos(+\psi_1 + \psi_2 - \rho + \chi),$$

$$\dot{\rho} = + \frac{\alpha}{2} \frac{r_2}{r_1} \cos(+\psi_1 + \psi_2 + \rho + \chi) - \frac{\alpha}{2} \frac{r_4}{r_1} \cos(+\psi_1 - \psi_2 + \rho + \chi)$$

$$- \frac{\alpha}{2} \frac{r_3}{r_2} \cos(+\psi_1 - \psi_2 - \rho + \chi) + \frac{\alpha}{2} \frac{r_1}{r_2} \cos(-\psi_1 - \psi_2 - \rho + \chi)$$

$$+ \frac{\alpha}{2} \frac{r_4}{r_3} \cos(-\psi_1 - \psi_2 + \rho + \chi) - \frac{\alpha}{2} \frac{r_2}{r_3} \cos(-\psi_1 + \psi_2 + \rho + \chi)$$

$$- \frac{\alpha}{2} \frac{r_1}{r_4} \cos(-\psi_1 + \psi_2 - \rho + \chi) + \frac{\alpha}{2} \frac{r_3}{r_4} \cos(+\psi_1 + \psi_2 - \rho + \chi). \quad (6.11)$$

Although this system appears to be more complicated than (6.7), its dimension is reduced by one. System (6.11) still possesses the D_4-symmetry inherited from the square-shaped coupling graph. We find the fundamental domain

$$(r_1, \ldots, r_4, \psi_1, \psi_2, \rho) \in \mathbb{R}_+^4 \times \left[-\frac{\pi}{2}, \frac{\pi}{2} \right]^3. \quad (6.12)$$

Appropriate reflections map the fundamental domain to the entire phase space. The decoupling subspace in the new coordinates,

$$D = \{ (r_1, \ldots, r_4, \psi_1, \psi_2, \rho) \mid \psi_1 = \psi_2 = 0 \ (\mathrm{mod} \ \pi), \ r_i = r_{i+2}, \ i = 1, \ldots, 4 \},$$

contains the line of equilibria

$$E = \{ (r_1, \ldots, r_4, \psi_1, \psi_2, \rho) \mid \psi_1 = \psi_2 = 0 \ (\mathrm{mod} \ \pi), \ r_i = 1, \ i = 1, \ldots, 4 \}. \quad (6.13)$$

Note how the line is parametrized by the phase angle $\rho = \arg z_1 - \arg z_2 - \frac{\pi}{2}$ between the two antipodal pairs.

On the fundamental domain, the $\mathbb{Z}_4$ symmetry

$$\begin{aligned} \delta &:= (r_1, r_2, r_3, r_4, \psi_1, \psi_2, \rho) \longmapsto (r_2, r_3, r_4, r_1, -\psi_2, \psi_1, -\rho), \\ \delta^2 &= (r_1, r_2, r_3, r_4, \psi_1, \psi_2, \rho) \longmapsto (r_3, r_4, r_1, r_2, -\psi_1, -\psi_2, \rho), \\ \delta^3 &= (r_1, r_2, r_3, r_4, \psi_1, \psi_2, \rho) \longmapsto (r_4, r_1, r_2, r_3, \psi_2, -\psi_1, -\rho) \end{aligned} \quad (6.14)$$

remains. Note in particular the reflection δ^2 which leaves the line of equilibria pointwise fixed.

We now linearize (6.11) at points on the line of equilibria. Under the assumption $f'(1) = -1$, see (6.7), we find four stable eigenvalues in radial directions and one trivial zero eigenvalue zero in direction of the line of equilibria. The two remaining eigenvalues determine the stability of the decoupled states on E. We find a zero

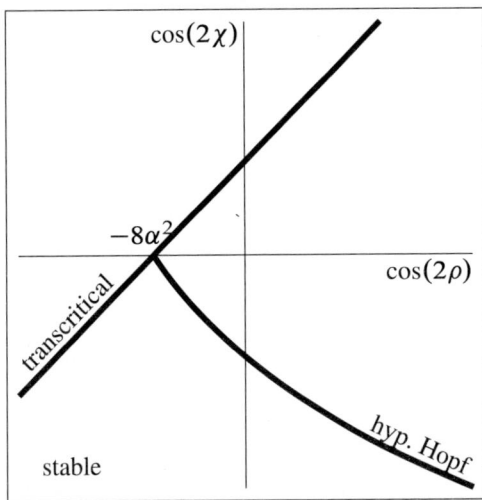

Fig. 6.2 Decoupling of a square ring of oscillators. Stability of the decoupled states depending on the phase angle ρ between antipodal pairs and the fixed coupling rotation angle χ: transcritical point: $\cos(2\rho) = \cos(2\chi) - 8\alpha^2$, (hyperbolic) Poincaré-Andronov Hopf point: $\cos(2\chi) < 0$ and $\cos(2\rho) = -\cos(2\chi) - 8\alpha^2(1 - \cos^2(2\chi))$

eigenvalue for $\cos(2\rho) = \cos(2\chi) - 8\alpha^2$ and a pair of purely imaginary eigenvalues for $\cos(2\chi) < 0$ and $\cos(2\rho) = -\cos(2\chi) - 8\alpha^2(1 - \cos^2(2\chi))$, see Fig. 6.2.

We have encountered $\mathbb{Z}_2$-symmetric transcritical bifurcations as well as hyperbolic Poincaré-Andronov-Hopf bifurcations, both without parameters. Transversality and non-degeneracy conditions are checked by a straight-forward calculation. Note that albeit χ is a classical parameter, ρ is not. Bifurcations occur for fixed χ along the line of equilibria parametrized by ρ.

In particular, at least a part of the decoupling subspace is normally stable: The decoupling is stabilized by a *non-invasive coupling*, that is a coupling that vanishes at the stabilized state.

Chapter 7
Application: Oscillatory Profiles in Systems of Hyperbolic Balance Laws

Conservation laws

$$u_t + F(u)_x = 0, \qquad x \in \mathbb{R}, \quad u \in \mathbb{R}^n, \tag{7.1}$$

or, more generally,

$$\frac{\partial}{\partial t} u + \sum_{i=1}^{k} \frac{\partial}{\partial x_i} f_i(u) = 0, \qquad x \in \mathbb{R}^k, \quad u \in \mathbb{R}^n, \tag{7.2}$$

arise in various physical models including fluid dynamics [43], magneto-hydro-dynamics [32], elasticity [44], multiphase flow in oil recovery [58], cosmology [68], and many more.

The prototype of a conservation law is the one-dimensional, scalar Burgers equation

$$\begin{aligned} u_t + (\tfrac{1}{2}u^2)_x &= 0 & \text{or, alternatively,} \\ u_t + u\,u_x &= 0, & x \in \mathbb{R}, u \in \mathbb{R}. \end{aligned} \tag{7.3}$$

It was introduced in [18] as a model of turbulence.

Shocks may form in finite time, and their admissibility as discontinuous solutions is a major question of the theory. One approach to this question are viscous regularizations of the conservation law. A typical requirement is the strict hyperbolicity of the system of conservation laws: all eigenvalues of F' should be distinct real numbers. This guarantees the well-posedness of the associated initial-value problem. For a more detailed introduction into the topic, see [22, 67].

Here we combine a strictly hyperbolic conservation law with a (stiff) source term.

$$u_t + f(u)_x = \tfrac{1}{\varepsilon} g(u), \qquad x \in \mathbb{R}, \quad u \in \mathbb{R}^n. \tag{7.4}$$

© Springer International Publishing Switzerland 2015
S. Liebscher, *Bifurcation without Parameters*, Lecture Notes in Mathematics 2117,
DOI 10.1007/978-3-319-10777-6_7

Both parts, alone, are "tame": The conservation law may form shocks, but in general stays piecewise smooth. Oscillatory tails of shocks may appear as numerical artifacts, only [47]. The source term, alone, will describe a simple, stable kinetic behavior: all trajectories eventually converge monotonically to some equilibrium. The balance law (7.4), constructed of these two parts, however, can support profiles with oscillatory tails. They emerge from Poincaré-Andronov-Hopf bifurcations without parameters in the associated traveling-wave system.

In [26,53], viscous profiles $u(t, x) = u((\xi - st)/\varepsilon)$ of the parabolic regularization

$$u_t + f(u)_x = \tfrac{1}{\varepsilon} g(u) + \varepsilon u_{xx} \tag{7.5}$$

of (7.4) have been investigated. Viscous profiles satisfy the ε-independent ODE system

$$\ddot{u} = (f'(u) - s \cdot \mathrm{id})\dot{u} - g(u). \tag{7.6}$$

Standard conservation laws, for example, require $g \equiv 0$. The presence of m conservation laws corresponds to nonlinearities g with range in a manifold of dimension $n - m$ in u-space. Typically, then, $g(u) = 0$ describes an equilibrium manifold of dimension m of pairs $(u, \dot{u}) = (u, 0)$, in the phase space of (7.6). In [26, 53], Poincaré-Andronov-Hopf points, Chap. 5, and also Bogdanov-Takens point, Chap. 10, have been found along this manifold in particular examples. The result of [39] goes farther and holds for inviscid profiles and—by perturbation—for viscous profiles alike.

Theorem 7.1 ([39]) *Let $f : \mathbb{R}^3 \to \mathbb{R}^3$ be a generic C^6 vector field such that $Df(u)$ has only real distinct eigenvalues $\lambda_1(u) < \lambda_2(u) < \lambda_3(u)$ for all u in a neighborhood of the origin $u = 0$, i.e. the hyperbolic conservation law $u_t + f(u)_x = 0$ is strictly hyperbolic.*

Then, for every value $s \notin \{\lambda_1(0), \lambda_2(0), \lambda_3(0)\}$ there exists a C^5 vector field

$$g : \mathbb{R}^3 \to \mathbb{R}^2 \times \{0\} \tag{7.7}$$

such that

(i) the kinetic part g stabilizes the line of equilibria near the origin, i.e. the linearization $Dg(0)$ has one (trivial) zero eigenvalue and two negative real eigenvalues,

(ii) the traveling-wave equation

$$u' = (Df(u) - s \cdot \mathrm{id})^{-1} g(u). \tag{7.8}$$

admits a Poincaré-Andronov-Hopf point in the sense of Chap. 5.

Proof Without loss of generality, we assume all eigenvalues of $Df(0)$ to be nonzero and $s = 0$. We shall construct a particular source g with a straight line of equilibria.

The construction starts with a suitable linearization $Dg(0)$ which creates the purely imaginary eigenvalues of $Df(0)^{-1}Dg(0)$. Then, this linearization is continued along the line of equilibria such that the transversality condition (5.7) holds. Finally, the non-degeneracy condition (5.8) is translated into a non-degeneracy condition on f. The main obstruction in the construction is the constraint (7.7) imposed by the structure of one conservation law and two balance laws.

Let S be the transformation of $Df(0)$ into diagonal form:

$$Df(0) = S \Lambda S^{-1}, \qquad \Lambda = \operatorname{diag}(\lambda_1, \lambda_2, \lambda_3). \qquad (7.9)$$

A Hopf point of system (7.8) at the origin requires the existence of a transformation $T \in GL(3)$, such that

$$Dg(0) = S \Lambda S^{-1} T^{-1} \begin{pmatrix} 0 & 0 & 0 \\ 0 & 0 & 1 \\ 0 & -1 & 0 \end{pmatrix} T. \qquad (7.10)$$

Here we have normalized the imaginary part of the Hopf eigenvalue to one. Indeed T is then the linearized normal-form transformation of (7.8) onto Hopf normal form. On this linear level, the constraint (7.7) yields $0 = e_3^T Dg(0)$ which is equivalent to

$$\Lambda S^T e_3 \perp S^{-1} T^{-1} \left(\{0\} \times \mathbb{R}^2 \right), \qquad (7.11)$$

where $e_3 = (0, 0, 1)^T$ denotes the third standard unit vector.

Aside from (7.11) we can choose T freely in order to construct the two negative eigenvalues of $Dg(0)$ defined by (7.10). To do this, we start with two arbitrary, linearly independent vectors $(a_1, a_3), (a_2, a_4) \in \mathbb{R}^2$. (The actual choice of the coefficients will be made later.) Under the first genericity condition of f,

$$e_3^T S e_k \neq 0, \qquad k = 1, 2, 3, \qquad (7.12)$$

we can choose a basis of $\mathbb{R}^3$ by:

$$v_1 := \begin{pmatrix} 1 \\ 0 \\ 0 \end{pmatrix}, \quad v_2 := \begin{pmatrix} * \\ a_1 \\ a_3 \end{pmatrix}, \quad v_3 := \begin{pmatrix} * \\ a_2 \\ a_4 \end{pmatrix}, \quad v_2, v_3 \in (\Lambda S^T e_3)^{\perp}. \qquad (7.13)$$

Then we define T by the equation

$$(T S)^{-1} := \begin{pmatrix} v_1 & v_2 & v_3 \end{pmatrix} = \begin{pmatrix} 1 & * & * \\ 0 & a_1 & a_2 \\ 0 & a_3 & a_4 \end{pmatrix}. \qquad (7.14)$$

and insert it into (7.10) to obtain

$$
T\,Dg(0)\,T^{-1}
$$

$$
= T\,S\,\Lambda\,S^{-1}\,T^{-1}
\begin{pmatrix}
0 & 0 & 0\\
0 & 0 & 1\\
0 & -1 & 0
\end{pmatrix}
\tag{7.15}
$$

$$
= \frac{1}{a_1 a_4 - a_2 a_3}
\begin{pmatrix}
0 & * & *\\
0 & (\lambda_3 - \lambda_2)a_2 a_4 & \lambda_2 a_1 a_4 - \lambda_3 a_2 a_3\\
0 & \lambda_2 a_2 a_3 - \lambda_3 a_1 a_4 & (\lambda_3 - \lambda_2)a_1 a_3
\end{pmatrix}
$$

The lower right (2×2)-block has trace $(\lambda_3 - \lambda_2)(a_1 a_3 + a_2 a_4)/(a_1 a_4 - a_2 a_3)$ and determinant $\lambda_2 \lambda_3$. The trace can be made negative of arbitrary size regardless of λ_2, λ_3 by choice of $a_1, \ldots, a_4$. We conclude: if $\lambda_2 \lambda_3 > 0$ then we can find parameters $a_1, \ldots, a_4$ in (7.13) such that the resulting matrix $Dg(0)$ has two negative real eigenvalues. In fact, there is an open region of admissible parameters. The requirement $\lambda_2 \lambda_3 > 0$ can be fulfilled without loss of generality by a permutation of $\lambda_1, \lambda_2, \lambda_3$, since at least two of them must have the same sign.

Let v_1, v_2, v_3, T now be fixed according to the above considerations. This defines the linearization $Dg(0)$ consistent with our claim. Then we continue the critical eigendirection

$$
\ker Dg(0) = \operatorname{span}\left\{ S \begin{pmatrix} 1\\0\\0 \end{pmatrix} \right\} = \operatorname{span}\left\{ T^{-1} \begin{pmatrix} 1\\0\\0 \end{pmatrix} \right\}
\tag{7.16}
$$

to a straight line of equilibria by the choice

$$
\left(g \circ T^{-1} \right) \begin{pmatrix} w_1\\ w_2\\ w_3 \end{pmatrix} = S\,\Lambda\,S^{-1}\,T^{-1}
\begin{pmatrix}
0 & 0 & 0\\
0 & cw_1 & 1\\
0 & -1 & cw_1
\end{pmatrix}
\begin{pmatrix} w_1\\ w_2\\ w_3 \end{pmatrix}
\tag{7.17}
$$

The transversality of the Hopf eigenvalue (5.7) can be achieved by an appropriate choice of the parameter c dependent on the higher order terms of f. Again, there is in fact an open region of admissible parameter values.

The required nondegeneracy of the Hopf point (5.8) is the second nondegeneracy condition needed for f. Indeed, due to the constraint (7.7), the term

$$
Df(0)^{-1}\Delta_Z g(u)\big|_{u=0} \in Z = \operatorname{span}\{S v_2, S v_3\}.
\tag{7.18}
$$

does not contribute to (5.8). The value $P_0 \Delta_Z D f(0)^{-1} g(0)$ of the Hopf nondegeneracy condition is determined by first-order terms of g (that have been defined using $D f(0)$, only) and second-order terms of f. Thus, (5.8) directly translate to a (generic) condition on the second order terms of f. This completes the construction of g.   □

The non-degeneracy condition (5.8) is equivalent to the requirement that breakdown of every flow-invariant foliation transverse to the line of equilibria at the Hopf point is already caused by second order terms of the vector field. In terms of our system of conservation laws and balance laws, it requires in particular that the flux couples the components with source terms back to the pure conservation law. Without such a coupling, the conservation law would give rise to a foliation, such that in each fiber only finitely many of the equilibria remain.

Similar to Theorem 7.1, Bogdanov-Takens points can occur in systems with at least two conservation laws and two balance laws. An example is given in [27].

In summary, Poincaré-Andronov-Hopf points as well as Bogdanov-Takens points are possible in systems of stiff hyperbolic balance laws. For all generic, strictly hyperbolic flux functions and a suitable number of pure conservation laws and balance laws there exist appropriate source terms such that these bifurcations occur in a structurally stable fashion. The bifurcations are generated by the interaction of flux and source. In particular, Hopf points can be constructed for generic fluxes and stabilizing sources. This interaction of two individually stabilizing effects to create instabilities, oscillations, or patterns is similar in spirit to the Turing instability [71], although the Turing instability is caused by the interaction of a stable kinetics with diffusion instead of transport.

This holds true under small perturbations of the system, for instance in numerical calculations. In particular, an additional viscous regularization

$$u_t + f(u)_x = g(u) + \delta u_{xx} \tag{7.19}$$

still yields the bifurcation scenario for small positive δ.

Note that traveling waves corresponding to heteroclinic orbits near an elliptic Hopf point have oscillatory tails, see Fig. 7.1. Hyperbolic conservation laws are usually expected to have monotone viscous shock profiles. In particular, in numerical simulations small oscillations near the shock layer are regarded as numerical artifacts due to grid phenomena or unstable numerical schemes. In many schemes "artificial viscosity" is used to automatically suppress such oscillations as "spurious". However, near elliptic Hopf points, all heteroclinic orbits correspond to traveling waves with necessarily oscillatory tails. Numerical schemes should therefore resolve this "overshoot" rather than suppress it.

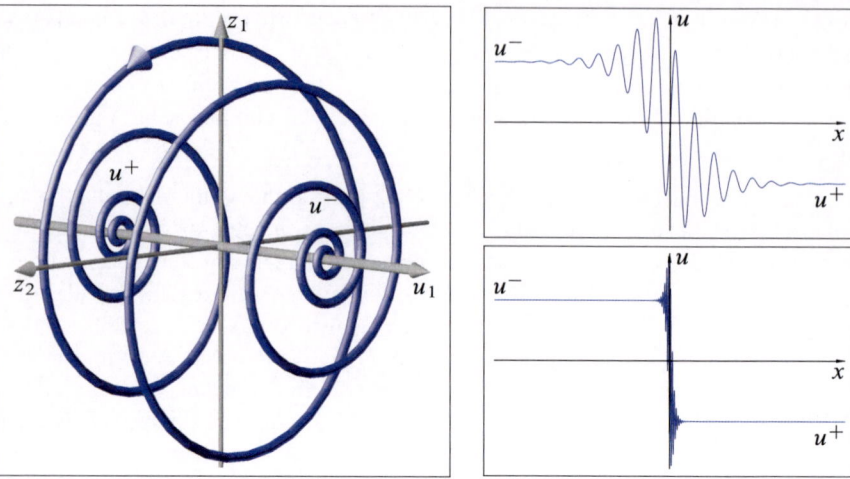

Fig. 7.1 Oscillatory profile near elliptic Poincaré-Andronov-Hopf point. *Left*: Heteroclinic orbit near the elliptic Poincaré-Andronov-Hopf point without parameter. *Right*: Profile for decreasing values of ε of the balance-law systems (7.5) or (7.4)

Additionally, in [53] convective stability of the resulting traveling waves has been proved, if their speed is large enough. For numerical calculations on bounded intervals in co-moving coordinates this implies nonlinear stability of the corresponding oscillatory traveling waves, as long as no artificial instabilities are introduced by inadequate boundary conditions.

Part III
Codimension Two

Chapter 8
Degenerate Transcritical Bifurcation

Along two-dimensional equilibrium manifolds, we expect transcritical points, Chap. 4, to form one-dimensional curves, by the implicit-function theorem. At isolated points, one of the non-degeneracy conditions (4.8, 4.9) may fail and codimension-two singularities appear. We shall discuss these degeneracies, first in a one-parameter-family of lines of equilibria and then along a two-dimensional equilibrium surface.

8.1 Families of Lines of Equilibria: Singular Drift

With one parameter, the degeneracy of the drift condition (4.9) is put as follows. We consider a system

$$
\begin{aligned}
\begin{pmatrix} \dot{x} \\ \dot{y} \end{pmatrix} &= F(x, y, \lambda) = \begin{pmatrix} f(x, y, \lambda) \\ g(x, y, \lambda) \end{pmatrix} \\
\dot{\lambda} &= 0,
\end{aligned}
\left.\rule{0pt}{40pt}\right\} \quad x, y, \lambda \in \mathbb{R}, \quad (8.1)
$$

with the following properties:

(i) For all parameter values, there exists a line of equilibria, $F(0, y, \lambda) \equiv 0$, forming a plane of equilibria in the extended phase space.
(ii) For all parameter values, the origin is a transcritical point, i.e. the origin has an eigenvalue zero in transverse direction to the equilibrium plane, $\partial_x f(0, 0, \lambda) \equiv 0$.
(iii) For all parameter values, this nontrivial eigenvalue crosses zero with nonvanishing speed as y increases, $\partial_y \partial_x f(0, 0, 0) > 0$.

© Springer International Publishing Switzerland 2015
S. Liebscher, *Bifurcation without Parameters*, Lecture Notes in Mathematics 2117,
DOI 10.1007/978-3-319-10777-6_8

(iv) At $\lambda = 0$ the drift non-degeneracy condition fails, $\partial_x g(0,0,0) = 0$.
 (v) This drift degeneracy is transverse, i.e. the drift changes direction with nonvanishing speed, as λ increases, $\partial_\lambda \partial_x g(0,0,0) > 0$.

The first condition is our structural assumption, (iii,v) are non-degeneracy assumptions which are fulfilled generically, and (ii,iv) describe our bifurcation point. Signs in (iii,v) are chosen without loss of generality, by switching signs of y and λ, if necessary.

Instead of (ii) it suffices to require $\partial_x f(0,0,0) = 0$ at the origin, only. Then the implicit-function theorem together with (iii) yields $\partial_x f(0, y(\lambda), \lambda) \equiv 0$ along a curve. Without loss of generality, we took this curve to be the λ-axis.

This setup is robust, i.e. under small perturbations of F respecting (i) there is a point near the origin satisfying (ii–v) for the perturbed system. From the point of view of singularity theory, (ii,iv) define a singularity of codimension two, which is unfolded versally by the coordinate y along the line of trivial equilibria and the parameter λ.

Condition (i) allows us to factor out x,

$$F(x, y, \lambda) = x\tilde{F}(x, y, \lambda), \tag{8.2}$$

with smooth $\tilde{F}$. Conditions (ii–v) yield an expansion

$$\tilde{F}(x, y, \lambda) = \begin{pmatrix} ax + by \\ cx + dy + \sigma\lambda \end{pmatrix} + \mathcal{O}((|x| + |y| + |\lambda|)^2), \tag{8.3}$$

with coefficients $a, b, c, d, \sigma \in \mathbb{R}, b > 0, \sigma > 0$. We assume an additional non-degeneracy condition

(vi) The matrix

$$\partial_{(x,y)} \left(\frac{1}{x} F \right) (0,0,0) = \begin{pmatrix} a & b \\ c & d \end{pmatrix}$$

   is hyperbolic, i.e. has no purely imaginary eigenvalues.

Setting

$$\delta := ad - bc, \qquad \tau := a + d, \tag{8.4}$$

for determinant and trace, we therefore have $\delta \neq 0$, and $\tau \neq 0$ if $\delta > 0$.

Applying the multiplier x^{-1} to system (8.2) preserves trajectories for $x \neq 0$ but reverses their direction for $x < 0$. After the coordinate transformation $\tilde{x} = x$, $\tilde{y} = ax + by$, $\tilde{\lambda} = b\sigma\lambda$, we obtain

$$\begin{pmatrix} \tilde{x}' \\ \tilde{y}' \end{pmatrix} = \begin{pmatrix} \tilde{y} \\ -\delta\tilde{x} + \tau\tilde{y} + \tilde{\lambda} \end{pmatrix} + \mathcal{O}((|x| + |y| + |\lambda|)^2). \tag{8.5}$$

This yields a bifurcating equilibrium at $(\tilde{x}, \tilde{y}) \approx (\tilde{\lambda}/\delta, 0)$. Transversality of the branch of equilibria with respect to the trivial line of equilibria as well as the hyperbolicity of the nontrivial equilibria is ensured by condition (vi). Therefore, terms of higher order in (8.5) will preserve this structure. See Fig. 8.1 for phase portraits in various cases. Note the appearance of the generic transcritical bifurcation without parameters, Fig. 4.1, for $\lambda \neq 0$.

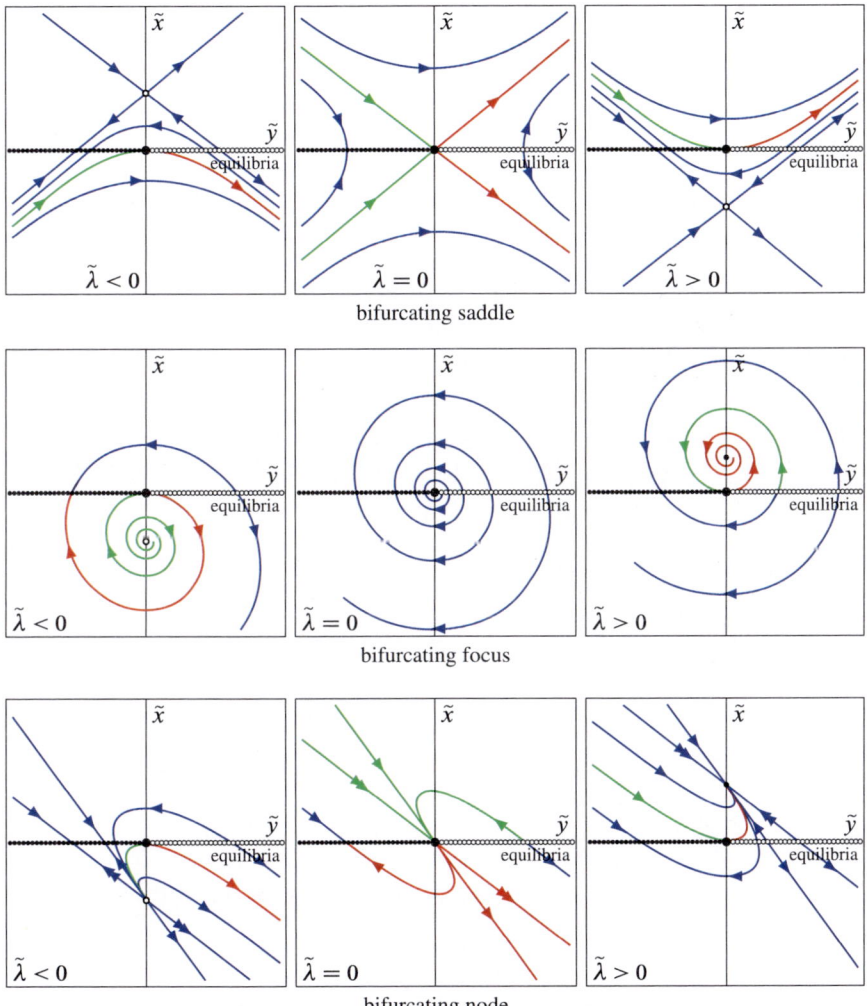

Fig. 8.1 Drift singularity along a one-parameter family of transcritical points. Stable set of the origin in *green*, unstable set in *red*

8.2   Families of Lines of Equilibria: Fold

With one parameter, also the transversality condition (4.8) could fail. We consider a system

$$\begin{pmatrix} \dot{x} \\ \dot{y} \end{pmatrix} = F(x, y, \lambda) = \begin{pmatrix} f(x, y, \lambda) \\ g(x, y, \lambda) \end{pmatrix} \left.\begin{matrix} \\ \\ \\ \end{matrix}\right\} \quad x, y, \lambda \in \mathbb{R}, \qquad (8.6)$$
$$\dot{\lambda} \;\; = 0,$$

with the following properties:

(i) For all parameter values, there exists a line of equilibria, $F(0, y, \lambda) \equiv 0$, forming a plane of equilibria in the extended phase space.

(ii) The origin is a transcritical point, i.e. the linearization at the origin has an eigenvalue zero in transverse direction to the equilibrium plane, $\partial_x f(0, 0, 0) = 0$.

(iii) Transversality of the eigenvalue, as y increases, fails: $\partial_y \partial_x f(0, 0, 0) = 0$.

(iv) The non-transversality in (iii) is unfolded versally, that is $\partial_\lambda \partial_x f(0, 0, 0) > 0$ and $\partial_y^2 \partial_x f(0, 0, 0) < 0$.

(v) The drift does not vanish, $\partial_x g(0, 0, 0) > 0$.

Again, the first condition is our structural assumption, (iv,v) are non-degeneracy assumptions which are fulfilled generically, and (ii,iii) describe our bifurcation point. The signs in (iv,v) are chosen without loss of generality, by switching signs of time, y, and λ, is necessary.

Note how (iii,iv) and the implicit-function theorem yield a fold-shaped curve of transcritical equilibria $(0, y, \lambda(y))$, two for each $\lambda > 0$ and none for $\lambda < 0$. Again, the setup is robust.

After factoring out x,

$$F(x, y, \lambda) \;=\; x\tilde{F}(x, y, \lambda) \;=\; x \begin{pmatrix} \tilde{f}(x, y, \lambda) \\ \tilde{g}(x, y, \lambda) \end{pmatrix}, \qquad (8.7)$$

we find the system

$$\dot{x} \;=\; \tilde{F}(x, y, \lambda), \qquad (8.8)$$

with the same orbits, outside the (y, λ)-plane of former equilibria, and reversed flow direction for $x < 0$.

Condition (v) implies $\tilde{g}(0, 0, 0) > 0$ and we can invoke the flow-box theorem and transform (8.8) to

$$\begin{aligned} \tilde{x}' &= 0, \\ \tilde{y}' &= 1. \end{aligned} \qquad (8.9)$$

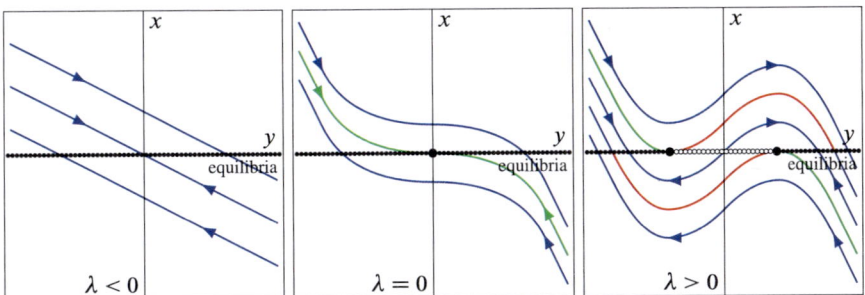

Fig. 8.2 Fold singularity along a one-parameter family of transcritical points. Stable set of the transcritical points in *green*, unstable set in *red*

Similar to Sect. 4.1, we determine the fate of the plane of equilibria by the implicit-function theorem. It is transformed to the family of curves

$$\tilde{x} = p(\tilde{y}, \lambda) = a\tilde{y}^3 + b\lambda\tilde{y} + \mathcal{O}(\tilde{y}^4, \tilde{y}^2\lambda, \tilde{y}\lambda^2), \tag{8.10}$$

with $a = -\frac{1}{6}\partial_y^2 \tilde{f}(0)\tilde{g}^2(0) > 0$ and $b = -\partial_\lambda \tilde{f}(0) < 0$, due to (iv, v). We can transform this, by another smooth coordinate change of y alone, to the standard cubic $\hat{x} = \frac{1}{3}\hat{y}^3 - \lambda\hat{y}$.

We arrive at the normal form

$$\begin{aligned} x' &= \lambda x - xy^2, \\ y' &= x, \end{aligned} \tag{8.11}$$

which has the same orbits and flow direction as (8.6) under a suitable coordinate transformation, see Fig. 8.2. We conclude:

Theorem 8.1 *Consider a $\mathscr{C}^4$ vector field (8.6) satisfying conditions (i–v). Then, there exists a $\mathscr{C}^1$-diffeomorphism which maps orbits of the vector field (8.6) to orbits of the normal form (8.11) with preserved time orientation.*

8.3   Planes of Equilibria

Along a plane of equilibria both singularities discussed in Sects. 8.1, 8.2 above turn out to be equivalent, see Remark 8.3.

Replacing the parameter λ discussed in Sect. 8.1 with an additional direction of a plane of equilibria, the drift along this manifold of equilibria is now a two-dimensional vector. It will not vanish along generic one-dimensional curves. The drift singularity along curves of transcritical points is therefore not characterized by a vanishing drift but rather by a drift direction tangential to the curve of transcritical points. (A drift in λ-direction was not possible in Sects. 8.1, 8.2 above.)

The correct setup is given by a system

$$\begin{pmatrix} \dot{x} \\ \dot{y} \end{pmatrix} = F(x, y) = \begin{pmatrix} f(x, y) \\ g(x, y) \end{pmatrix}, \qquad x \in \mathbb{R}, \quad y \in \mathbb{R}^2, \tag{8.12}$$

$y = (y_1, y_2)$, $g = (g_1, g_2)$, with the following properties:

(i) The y-plane consists of equilibria, $F(0, y) \equiv 0$.
(ii) There is a transcritical point at the origin, i.e. the y-plane loses normal hyperbolicity at this point, $\partial_x f(0,0) = 0$.
(iii) This loss of normal hyperbolicity is caused by the transverse eigenvalue crossing zero transversally, $\nabla_y \partial_x f(0,0) \neq 0$. Without loss of generality, the gradient points in y_1-direction, i.e. $\partial_{y_1} \partial_x f(0,0) > 0$ and $\partial_{y_2} \partial_x f(0,0) = 0$. By the implicit-function theorem, this gives rise to a curve of transcritical points tangential to the y_2-axis.
(iv) At the origin, the drift non-degeneracy transverse to the curve of transcritical points fails, $\partial_x g_1(0,0) = 0$.
(v) This drift degeneracy is transverse, i.e. the drift direction crosses the tangent to the curve of transcritical points with nonvanishing speed along the curve of transcritical points, $\partial_{y_1} \partial_x f(0,0)\, \partial_{y_2} \partial_x g_1(0,0) + \partial^2_{y_2} \partial_x f(0,0)\, \partial_x g_2(0,0) \neq 0$.
(vi) The drift does not vanish at the origin, i.e. there is a component tangential to the curve of transcritical points, $\partial_x g_2(0,0) > 0$.

Note that conditions (i–v) correspond to the conditions of Sect. 8.1. Again, the singularity described by (ii,iv) is robust under perturbations satisfying (i), provided the non-degeneracy conditions (iii,v,vi) hold. Signs in (iii,vi) are chosen without loss of generality, by switching signs of y_1 and y_2, if necessary.

The non-degeneracy condition (vi) indeed yields

$$\frac{d}{dy_2} \langle \nabla_y \partial_x f, \partial_x g \rangle (0, \vartheta(y_2), y_2) \Big|_{y_2=0} \neq 0, \tag{8.13}$$

where $(x, y_1, y_2) = (0, \vartheta(y_2), y_2)$, $\vartheta(0) = 0$, $\vartheta'(0) = 0$, is the curve γ of transcritical points. Locally, we could reparametrize y to achieve $\vartheta \equiv 0$. Conditions (iii,v) would then read: $\partial_x f(0,0,y_2) \equiv 0$, $\partial_{y_1} \partial_x f(0,0,0) > 0$, $\partial_{y_2} \partial_x g_1(0,0,0) \neq 0$. But let us continue with the original setup.

As in the parameter-dependent case (8.14), we can factor out x due to condition (i),

$$F(x, y) = x\tilde{F}(x, y) = x \begin{pmatrix} \tilde{f}(x, y) \\ \tilde{g}(x, y) \end{pmatrix}. \tag{8.14}$$

However, this time, due to non-degeneracy (vi) no equilibrium remains,

$$\tilde{F}(0,0,0) = (0, 0, \partial_x g_2(0,0,0)) \neq 0. \tag{8.15}$$

We apply the flow-box theorem: there exists a local smooth diffeomorphism

$$h(z_0, z_1, z_2) = \tilde{\Phi}_{z_2}(z_0, z_1, 0), \tag{8.16}$$

where $\tilde{\Phi}_t$ denotes the flow generated by the vector field $\tilde{F}$. This diffeomorphism fixes the origin and transforms $\tilde{F}$ into the constant vector field,

$$[Dh(z_0, z_1, z_2)]^{-1}\tilde{F}(h(z_0, z_1, z_2)) = \begin{pmatrix} 0 \\ 0 \\ 1 \end{pmatrix}. \tag{8.17}$$

Applying the same transformation to the original vector field F, we obtain

$$[Dh(z)]^{-1}F(h(z)) = [Dh(z)]^{-1}h_0(z)\tilde{F}(h(z)) = \begin{pmatrix} 0 \\ 0 \\ h_0(z) \end{pmatrix}, \tag{8.18}$$

where $h = (h_0, h_1, h_2)$.

In a suitable neighborhood of the origin, the vector field F is flow-equivalent to a vector field

$$\dot{z}_2 = h_0(z_0, z_1, z_2) \tag{8.19}$$

on the real line depending on two (classical) parameters (z_0, z_1). Expansion of h_0 using (8.16) and conditions (ii–vi) yields

$$\dot{z}_2 = az_2^3 + (c_0z_0 + c_1z_1)z_2^2 + (bz_1 + c_2z_0 + c_3z_0^2 + c_4z_0z_1 + c_5z_1^2)z_2 + z_0 + \mathcal{O}(|z|^4) \tag{8.20}$$

with

$$a = \left(\partial_{y_1}\partial_x f(0)\,\partial_{y_2}\partial_x g_1(0) + \partial_{y_2}^2\partial_x f(0)\,\partial_x g_2(0)\right)\partial_x g_2(0) \neq 0, \tag{8.21}$$
$$b = \partial_{y_1}\partial_x f(0) \neq 0.$$

In particular, $h_0(0, 0, z_2) = az_2^3 + \mathcal{O}(|z_2|^4)$. This is a cusp singularity. See [8,9,34,35, 62] for a background on singularity theory and its connection to dynamical systems. In fact, non-degeneracies (8.21) allow to diffeomorphically transform (8.20) into the normal form

$$\dot{z}_2 = \pm z_2^3 + z_1 z_2 + z_0 + \mathcal{O}(z_2^N), \tag{8.22}$$

for arbitrary normal-form order N, see for example [17], criterion 6.10. This is a minimal versal unfolding of the cusp singularity. See Fig. 8.3.

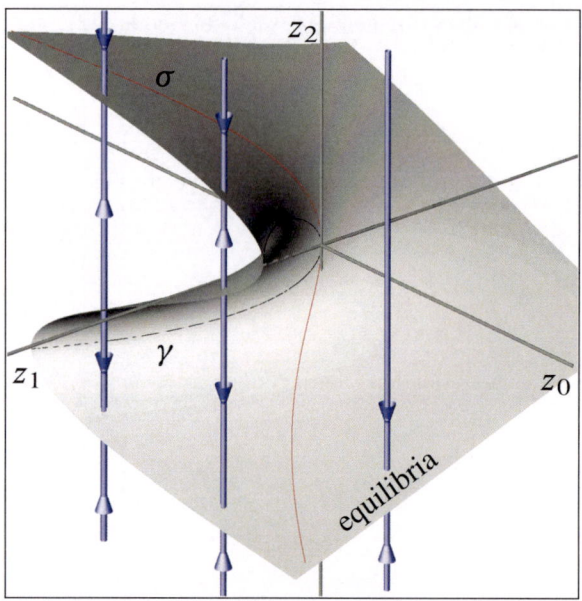

Fig. 8.3 Cusp singularity. Cusp singularity $\dot{z}_2 = az_2^3 + z_1z_2 + z_0$ with $a = -1$. Reverse direction of trajectories and signs of z_0, z_1 for $a = +1$. The fold line γ is connected by heteroclinic orbits to the curve σ, both curves have a common tangent at the origin

Reverting the flow-box transformation, the cusp singularity yields a description of the local dynamics near a transcritical point with drift singularity on a two-dimensional manifold of equilibria. Note in particular the cusp-shaped fold line

$$\gamma: \quad z_1^3 = \mp\frac{27}{4}z_0^2 + \mathcal{O}(z_0^{N/3}), \qquad z_2^3 = \pm\frac{1}{2}z_0 + \mathcal{O}(z_0^{N/3})$$

of the manifold of equilibria that is connected by heteroclinic orbits to the curve

$$\sigma: \quad z_1^3 = \mp\frac{27}{4}z_0^2 + \mathcal{O}(z_0^{N/3}), \qquad z_2^3 = \mp4z_0 + \mathcal{O}(z_0^{N/3}).$$

Theorem 8.2 *Under conditions (i–vi) the vector field (8.12) in a local neighborhood U of the origin is flow-equivalent to the cusp singularity (8.22). Depending on the sign of the cubic term*

$$a = \mathrm{sign}\left(\partial_{y_1}\partial_x f(0)\,\partial_{y_2}\partial_x g_1(0) + \partial_{y_2}^2\partial_x f(0)\,\partial_x g_2(0)\right),$$

all trajectories in U converge to an equilibrium $(0, y)$ in forward time $(a = -1)$ or backward time $(a = +1)$.

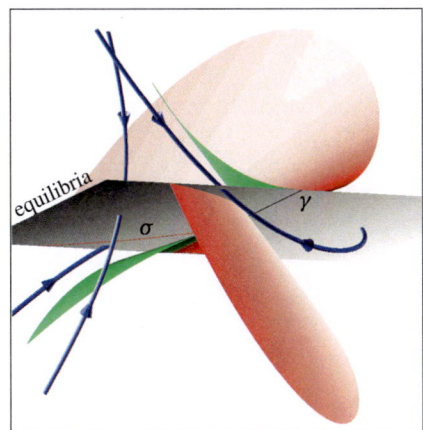 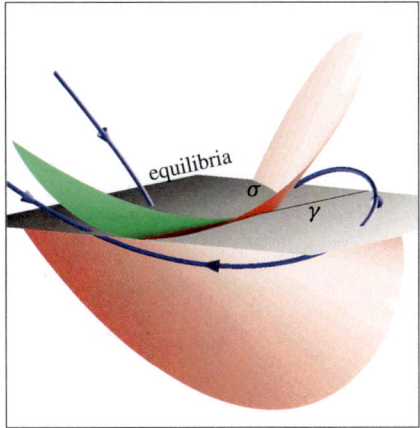

Fig. 8.4 Transcritical point with drift singularity on a plane of equilibria. Stable set of the line γ of transcritical points in *green*, unstable set in *red*, selected trajectories in *blue*. Two different views for $a = -1$. Reverse direction of trajectories and switch colors of manifolds for $a = +1$

In U, the transcritical points on the manifold of equilibria form a curve γ through the origin. The unstable (for $a = -1$) and stable (for $a = +1$) sets, respectively, of the two components γ_1, γ_2 of $\gamma \setminus \{0\}$ form manifolds of heteroclinic orbits on opposite sides of the manifold of equilibria. Their targets in forward time ($a = -1$) or backward time ($a = +1$) again form curves $\sigma_{1,2}$ on the manifold of equilibria with $\sigma_1 \cup \{0\} \cup \sigma_2$ being a tangential curve to γ. See Fig. 8.4.

Remark 8.3 There is no difference between drift and fold singularities, as discussed in Sects. 8.1 and 8.2 of one-parameter families of lines of equilibria, in the case of a plane of equilibria without parameters. Indeed, by a coordinate transformation of y, alone, the curve γ of transcritical points can be mapped onto the y_2-axis as well as onto a parabola tangential to the y_1 axis.

Remark 8.4 In contrast to the parameter-dependent drift singularity, equilibria do not bifurcate. In fact, the drift non-degeneracy excludes any kind of recurrent or stationary orbits except the primary manifold of equilibria.

Chapter 9
Degenerate Poincaré-Andronov-Hopf Bifurcation

In this chapter we study the Poincaré-Andronov-Hopf bifurcation without parameters, see Chap. 5, with an additional degeneracy of the drift or transversality due to an additional parameter or an additional dimension of the primary manifold of equilibria.

Compare the truncated normal forms of the classical Poincaré-Andronov-Hopf bifurcation

$$\begin{aligned} \dot{r} &= \lambda r + c r^3, \\ \dot{\varphi} &= 1. \end{aligned} \tag{9.1}$$

and the Poincaré-Andronov-Hopf point without parameters

$$\begin{aligned} \dot{r} &= r y, \\ \dot{y} &= \eta r^2, \\ \dot{\varphi} &= 1. \end{aligned} \tag{9.2}$$

See also (5.4), (5.11), both written in polar coordinates.

The classical bifurcation (9.1) can have two degeneracies: the coefficient of λr or the coefficient c of r^3 could vanish. Only the latter is usually called (classical) degenerate Hopf bifurcation. It features a transition from subcritical to supercritical Hopf bifurcation.

Without parameters, the drift $\dot{y} = \eta r^2$ distinguishes elliptic and hyperbolic cases. A transition between the two corresponds to a vanishing coefficient η. It will turn out, however, that on a plane of equilibria a degeneracy of the transversality, i.e. a vanishing coefficient of ry in (9.2), is equivalent to a drift degeneracy.

Again, we will consider the additional degeneracy to be induced by first varying a genuine parameter and then by considering a Poincaré-Andronov-Hopf on a two-dimensional surface of equilibria without parameters. It turns out that the mixed case of a one-parameter family of lines of equilibria and the pure case of a plane of equilibria yield different bifurcation scenarios.

© Springer International Publishing Switzerland 2015
S. Liebscher, *Bifurcation without Parameters*, Lecture Notes in Mathematics 2117,
DOI 10.1007/978-3-319-10777-6_9

9.1   Families of Lines of Equilibria: Singular Drift

Let us start with a Poincaré-Andronov-Hopf point on a curve of equilibria. Let
a degeneracy of the drift, see Chap. 5, be unfolded by one additional parameter.
Again, we restrict the problem to the now four-dimensional center manifold in a
neighborhood of the degenerate Poincaré-Andronov-Hopf point. In the restricted
system we deform the one-parameter family of curves of equilibria to a family of
straight lines. This yields the following setting: we consider a system

$$
\left.
\begin{aligned}
\begin{pmatrix} \dot{x} \\ \dot{y} \end{pmatrix} = F(x, y, \lambda) = \begin{pmatrix} f(x, y, \lambda) \\ g(x, y, \lambda) \end{pmatrix} \\
\dot{\lambda} \;\; = 0,
\end{aligned}
\right\} \qquad x \in \mathbb{R}^2, \quad y, \lambda \in \mathbb{R}, \qquad (9.3)
$$

$x = (x_1, x_2)$, $f = (f_1, f_2)$, with the following properties:

(i) For all parameter values, there exists a line of equilibria, $F(0, y, \lambda) \equiv 0$,
   forming a plane of equilibria in the extended phase space.
(ii) For all parameter values, the origin is an Andronov-Hopf point, i.e. the origin,
   has a pair of purely imaginary, nonzero eigenvalues in transverse direction to
   the equilibrium plane:

$$
\partial_x f(0, 0, \lambda) \equiv \begin{pmatrix} 0 & -1 \\ 1 & 0 \end{pmatrix}.
$$

(iii) For all parameter values, this nontrivial eigenvalue pair crosses the imaginary
   axis with nonvanishing speed as y increases, $\partial_y \mathrm{div}_x f(0, 0, \lambda) > 0$.
(iv) At $\lambda = 0$, the drift non-degeneracy condition fails, $\Delta_x g(0, 0, 0) = 0$, with
   $\Delta_x = \partial_{x_1}^2 + \partial_{x_2}^2$.
(v) This drift degeneracy is transverse, i.e. the drift changes direction with
   nonvanishing speed, as λ increases, $\partial_\lambda \Delta_x g(0, 0, 0) > 0$.

The first condition is our structural assumption, (iii,v) are non-degeneracy assumptions fulfilled generically, and (ii,iv) describe our bifurcation point. Note that we
chose coordinates $x = (x_1, x_2)$ such that $\partial_x f(0, 0, \lambda)$ is in Jordan normal form. We
further normalized the critical eigenvalue to $\pm i$. This can always be achieved by a
λ-dependent time rescaling, i.e. a scalar multiplier to the system, which preserves
the trajectories of the system. This setup is robust, i.e. under small perturbations of
F respecting (i) there is a point near the origin satisfying (ii–v) for the perturbed
system. From the point of view of singularity theory, (ii,iv) define a singularity of
codimension two, which is unfolded versally by the coordinate y along the line of
trivial equilibria and the parameter λ.

The linearization

$$A = DF(0,0,0) = \begin{pmatrix} 0 & -1 & 0 \\ 1 & 0 & 0 \\ 0 & 0 & 0 \end{pmatrix}$$

at the origin is normal. Thus, the normal-form procedure, see Sects. 2.2, 2.3, and [72], yields a normal form which is equivariant with respect to the group of rotations $\{\exp(DF(0,0,0)^{\mathsf{T}}\tau)\ ;\ \tau \in \mathbb{R}\}$. Writing (x_1, x_2) in polar coordinates, $(x_1 + ix_2) = r \exp(i\varphi)$, we obtain

$$
\begin{aligned}
\dot{r} &= rh_r(r^2, y, \lambda) + \text{h.o.t.,} \\
\dot{\varphi} &= h_\varphi(r^2, y, \lambda) + \text{h.o.t.,} \\
\dot{y} &= h_y(r^2, y, \lambda) + \text{h.o.t.,} \\
\dot{\lambda} &= 0.
\end{aligned}
\tag{9.4}
$$

with polynomials h_r, h_φ, h_y, in normal form, depending on r^2, y, λ but not on the angle φ. The terms of higher order, not in normal form, depend on all variables r, φ, y, λ and generically break the normal-form symmetry.

The plane of equilibria and the linearization at the origin remain unchanged by the normal form procedure, thus $h_r(0,0,0) = 0$, $h_\varphi(0,0,0) = 1$, $h_y(0, y, \lambda) \equiv 0$, due to conditions (i) and (ii). The multiplier $1/\dot{\varphi}$ is close to 1, preserves trajectories, and normalizes the rotation speed. Thus we can put $\dot{\varphi} = 1$ in (9.4). Condition (iii) translates to $\partial_y h_r(0,0,0) > 0$. We can introduce a new variable $\tilde{y} = h_r(0, y, \lambda)$ and obtain

$$
\begin{aligned}
\dot{r} &= r\tilde{y} + r^3\tilde{h}_r(r^2, \tilde{y}, \lambda) + \text{h.o.t.,} \\
\dot{\varphi} &= 1, \\
\dot{\tilde{y}} &= r^2\tilde{h}_y(r^2, \tilde{y}, \lambda) + \text{h.o.t.,} \\
\dot{\lambda} &= 0.
\end{aligned}
\tag{9.5}
$$

Finally, the drift-degeneracy conditions (iv) and (v) imply $\tilde{h}_y(0,0,0) = 0$ and $\partial_\lambda \tilde{h}_y(0,0,0) \neq 0$. We set $\tilde{\lambda} = \tilde{h}_y(0,0,\lambda)$ and obtain

$$
\begin{aligned}
\dot{r} &= r\tilde{y} + r^3\tilde{\tilde{h}}_r(r^2, \tilde{y}, \tilde{\lambda}) + \text{h.o.t.,} \\
\dot{\varphi} &= 1, \\
\dot{\tilde{y}} &= \lambda r^2 + r^2\tilde{\tilde{h}}_y(r^2, \tilde{y}, \tilde{\lambda}) + \text{h.o.t.,} \\
\dot{\lambda} &= 0,
\end{aligned}
\tag{9.6}
$$

with $\tilde{\tilde{h}}_y(0,0,\tilde{\lambda}) \equiv 0$.

Dropping tildes to simplify the notation, and omitting the trivial φ and λ directions as well as terms beyond normal form order, we obtain the (truncated) normal form

$$
\begin{aligned}
\dot{r} &= ry + r^3 h_r(r^2, y, \lambda), \\
\dot{y} &= \lambda r^2 + r^2 h_y(r^2, y, \lambda),
\end{aligned}
\tag{9.7}
$$

still with $h_y(0, 0, \lambda) \equiv 0$.

As we are interested in the local dynamics close to the origin, we rescale this system by

$$
\begin{aligned}
r &= \sigma \tilde{r}, \\
y &= \sigma^2 \tilde{y}, \\
\lambda &= \sigma^2 \tilde{\lambda}, \\
t &= \sigma^{-1} \tilde{t}.
\end{aligned}
\tag{9.8}
$$

For $0 < \sigma \ll 1$, to leading order in σ we obtain the rescaled, truncated normal form

$$
\begin{aligned}
\dot{r} &= ry + r\mathcal{O}(\sigma), \\
\dot{y} &= \lambda r^2 + \varrho_1 y r^2 + \varrho_2 r^4 + r^2 \mathcal{O}(\sigma).
\end{aligned}
\tag{9.9}
$$

If $\varrho_1 < 0$ then we switch its sign by replacing $(y, t) \mapsto -(y, t)$. If $\varrho_1 > 0$, then we can normalize it to $\varrho_1 = 1$ by scaling r, λ. To ensure $\varrho_1 \neq 0$, we assume the additional non-degeneracy condition:

(vi) $0 \neq \partial_y \left(\partial_{x_1}^2 + \partial_{x_2}^2\right) g(0, 0, 0) = \partial_y \Delta_x g(0, 0, 0)$.

Then we obtain to leading order in the rescaling parameter σ:

$$
\begin{aligned}
\dot{r} &= ry, \\
\dot{y} &= \lambda r^2 + y r^2 + \varrho r^4.
\end{aligned}
\tag{9.10}
$$

To this system, we apply the multiplier $1/r$. Trajectories in the domain $\{r > 0\}$ are preserved. The boundary $\{r = 0\}$ still represents the line of equilibria. We arrive at the truncated normal form

$$
\begin{aligned}
r' &= y, \\
y' &= \lambda r + yr + \varrho r^3.
\end{aligned}
\tag{9.11}
$$

Note the reversibility with respect to the reflection $r \mapsto -r$ induced by the normal-form symmetry, that is the independence of (9.7) of the angle φ. The equilibria of (9.11) are the origin, $(r, y) = (0, 0)$, and the points $(r, y) = (\pm\sqrt{-\lambda/\varrho}, 0)$.

Of course, for $\lambda \neq 0$, we find a generic Poincaré-Andronov-Hopf bifurcation without parameters at the origin: compare our setting (9.3) with Chap. 5, in particular the truncated normal forms (9.10) and (5.11). We find the elliptic case for $\lambda < 0$ and the hyperbolic case for $\lambda > 0$.

The pair of bifurcating equilibria

$$\begin{pmatrix} r \\ y \end{pmatrix} = \begin{pmatrix} \pm\sqrt{-\lambda/\varrho} \\ 0 \end{pmatrix}, \quad \text{with linearization} \quad \begin{pmatrix} 0 & 1 \\ -2\lambda & \pm\sqrt{-\lambda/\varrho} \end{pmatrix} \quad (9.12)$$

accompany the elliptic or the hyperbolic Hopf points depending on the sign of ϱ. It corresponds to a periodic orbit of the full system (9.6). Indeed, for $\lambda \neq 0$ the equilibria (9.12) are hyperbolic, i.e. their linearizations have no purely imaginary eigenvalues. They are also hyperbolic fixed points of the time-2π map to the vector field (9.10). Thus, they persist under small perturbations. This yields a hyperbolic equilibrium of the Poincaré map to the full system (9.6): the claimed hyperbolic periodic orbit.

For $\varrho > 0$, we find the bifurcating equilibria (9.12) for $\lambda < 0$ accompanying the elliptic Hopf points. The determinant of the linearization is negative: we find a saddle periodic orbit, its stable/unstable manifolds bound the elliptic bubble of bounded trajectories. See Fig. 9.1. We call this the subcritical case.

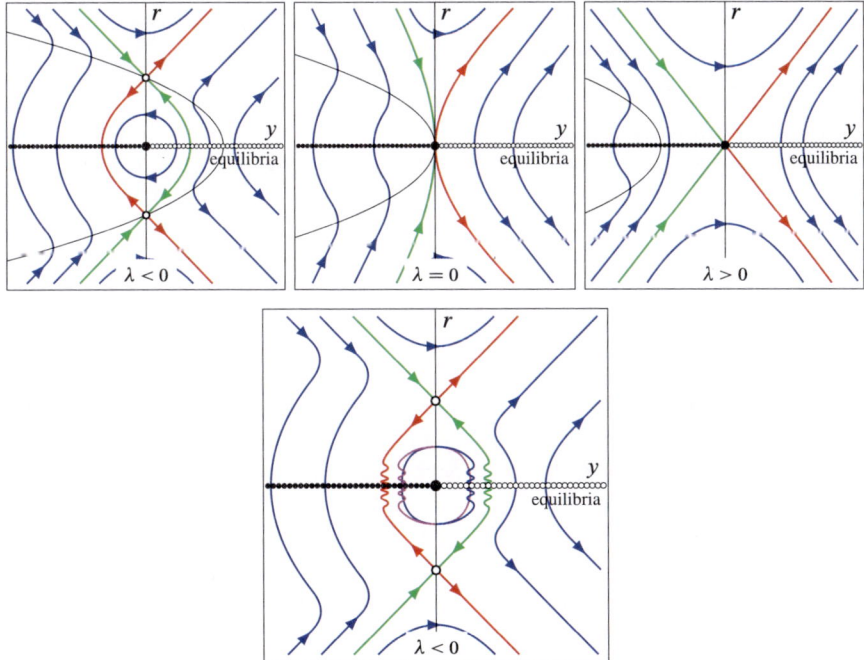

Fig. 9.1 Andronov-Hopf point with drift singularity, subcritical case. *Upper part*: Normal form flow near an Andronov-Hopf with drift singularity alias drift singularity of pitchfork bifurcations; stable set of the origin or the bifurcating saddle in *green*, unstable set in *red*; nullclines in *black*. Note the elliptic Hopf point, for $\lambda < 0$, and the hyperbolic Hopf point, for $\lambda > 0$. *Lower part*: Poincaré map with separatrix splitting; compare with the left picture above

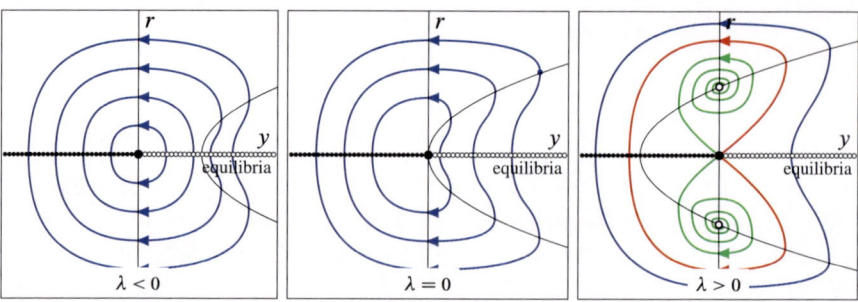

Fig. 9.2 Andronov-Hopf point with drift singularity, supercritical case. Normal form flow near an Andronov-Hopf with drift singularity alias drift singularity of pitchfork bifurcations; stable set of the origin in *green*, unstable set in *red*; nullclines in *black*. Note the elliptic Hopf point, for $\lambda < 0$, and the hyperbolic Hopf point, for $\lambda > 0$

For $\varrho < 0$, we find the bifurcating equilibria (9.12) for $\lambda > 0$ accompanying the hyperbolic Hopf points. The determinant of the linearization is negative: we find a periodic orbit inside the elliptic bubble of bounded trajectories. The periodic orbit is of node type, i.e. with real Floquet exponents, for $|\varrho| \leq 1/8$, and of focus type for $|\varrho| > 1/8$. Due to our choice of sign of ϱ_1 in (9.9), it is unstable. All trajectories close to the origin remain bounded. See Fig. 9.2. We call this the supercritical case.

Theorem 9.1 *Under the conditions (i–v, vi) the vector field (9.3) in rescaled polar coordinates (and possibly under time reversal) has the truncated normal form*

$$\dot{r} = ry,$$
$$\dot{\varphi} = 1,$$
$$\dot{y} = \lambda r^2 + yr^2 + \varrho r^4.$$

In the subcritical case, $\varrho > 0$, non-trivial bounded local trajectories exist for $\lambda < 0$. The set of bounded trajectories is formed by heteroclinic orbits to trivial equilibria on opposite sides of the elliptic Hopf point at the origin. It is bounded by the stable and unstable manifolds to the bifurcating saddle periodic orbit.

In the supercritical case, $\varrho < 0$, all local trajectories are bounded. For $\lambda \leq 0$, the set of bounded trajectories is formed by heteroclinic orbits to trivial equilibria on opposite sides of the elliptic Hopf point at the origin. For $\lambda > 0$, additional heteroclinic orbits connect the bifurcating unstable (or stable) periodic orbit to the trivial equilibria on one side of the hyperbolic Hopf point.

Terms of higher order in (9.5) beyond normal form depend on φ and will cause the separatrices to split, as already observed close to the generic Hopf point, Chap. 5. In addition to the splitting of two-dimensional stable/unstable manifolds of primary equilibria, we find that the stable/unstable manifold of the bifurcating periodic orbit in the subcritical case connects to a small interval of primary equilibria. For analytic vector fields, Neishtadt averaging yields exponential smallness of this splitting with respect to the distance from the origin, that is exponential smallness in λ.

9.2 Families of Lines of Equilibria: Fold

As we have done in Sect. 8.2 for the transcritical point, we study the failure of the transversality of the eigenvalue crossing at the Andronov-Hopf point. Let us again start with an Andronov-Hopf point on a curve of equilibria. On the center manifold, we consider a system

$$
\begin{pmatrix} \dot{x} \\ \dot{y} \end{pmatrix} = F(x, y, \lambda) = \begin{pmatrix} f(x, y, \lambda) \\ g(x, y, \lambda) \end{pmatrix} \left. \right\} \quad x \in \mathbb{R}^2, \quad y, \lambda \in \mathbb{R}, \qquad (9.13)
$$
$$
\dot{\lambda} = 0,
$$

$x = (x_1, x_2)$, $f = (f_1, f_2)$, with the following properties:

 (i) For all parameter values, there exists a line of equilibria, $F(0, y, \lambda) \equiv 0$, forming a plane of equilibria in the extended phase space.
 (ii) The origin is a Poincaré-Andronov-Hopf point, i.e. the origin has pair of purely imaginary, nonzero eigenvalues in transverse direction to the equilibrium plane:

$$
\partial_x f(0, 0, 0) = \begin{pmatrix} 0 & -1 \\ 1 & 0 \end{pmatrix}.
$$

(iii) Transversality of the purely imaginary eigenvalue pair, as y increases, fails at the origin: $0 = \partial_y \mathrm{div}_x f(0, 0, 0) = \partial_y (\partial_{x_1} f_1(0) + \partial_{x_2} f_2(0))$.
(iv) The non-transversality in (iii) is unfolded versally, that is $\partial_\lambda \mathrm{div}_x f(0, 0, 0) > 0$ and $\partial_y^2 \mathrm{div}_x f(0, 0, 0) < 0$.
 (v) The drift along the line of equilibria is non-degenerate, i.e. $\Delta_x g(0, 0, 0) > 0$, with $\Delta_x = \partial_{x_1}^2 + \partial_{x_2}^2$.

The first condition is our structural assumption, (iv,v) are non-degeneracy assumptions fulfilled generically, and (ii,iii) describe our bifurcation point. Signs in (iii,v) are chosen without loss of generality, by switching signs of time, y, and λ, if necessary.

Note how (iii,iv) and the implicit-function theorem again yield a fold-shaped curve of Poincaré-Andronov-Hopf points $(0, y, \lambda(y))$, two for each $\lambda > 0$ and none for $\lambda < 0$. Again, the setup is robust.

In analogy to the previous section, we find a normal form with additional rotationally symmetry,

$$
\begin{aligned}
\dot{r} &= r h_r(r^2, y, \lambda) + \text{h.o.t.,} \\
\dot{\varphi} &= 1, \\
\dot{y} &= r^2 h_y(r^2, y, \lambda) + \text{h.o.t.,} \\
\dot{\lambda} &= 0.
\end{aligned} \qquad (9.14)
$$

The conditions (iii–v) are equivalent to $h_r(0) = 0$, $\partial_\lambda h_r(0) > 0$, $\partial_y^2 h_r(0) < 0$, and $h_y(0) > 0$. We set $\tilde\lambda = h_r(0,0,\lambda)$ and $\tilde y^2 = h_r(0,0,\lambda) - h_r(0,y,\lambda)$ to obtain

$$\begin{aligned}
\dot r &= r(\tilde\lambda - \tilde y^2) + r^3 \tilde h_r(r^2, \tilde y, \tilde\lambda) + \text{h.o.t.},\\
\dot\varphi &= 1,\\
\dot{\tilde y} &= cr^2 + r^2 \tilde h_y(r^2, \tilde y, \tilde\lambda) + \text{h.o.t.},\\
\dot{\tilde\lambda} &= 0,
\end{aligned}\tag{9.15}$$

with $\tilde h_y(0,0,0) = 0$ and a constant $c > 0$ which can be normalized to 1 by scaling of r.

Dropping tildes to simplify the notation, and omitting the trivial φ and λ directions as well as terms beyond normal form order, we obtain the (truncated) normal form

$$\begin{aligned}
\dot r &= r(\lambda - y^2) + r^3 h_r(r^2, y, \lambda),\\
\dot y &= r^2 + r^2 h_y(r^2, y, \lambda),
\end{aligned}\tag{9.16}$$

still with $h_y(0,0,0) = 0$.

We rescale this system by

$$\begin{aligned}
r &= \sigma^3 \tilde r,\\
y &= \sigma^2 \tilde y,\\
\lambda &= \sigma^4 \tilde\lambda,\\
t &= \sigma^{-4} \tilde t.
\end{aligned}\tag{9.17}$$

For $0 < \sigma \ll 1$, to leading order in σ we obtain the rescaled, truncated normal form

$$\begin{aligned}
\dot r &= r\lambda - ry^2 + r\mathcal{O}(\sigma),\\
\dot y &= r^2 + r^2 \mathcal{O}(\sigma).
\end{aligned}\tag{9.18}$$

Theorem 9.2 *Assume (i–v) for the vector field (9.13). Then, for $\lambda < 0$ we find a hyperbolic Andronov-Hopf point at $y \approx -\sqrt\lambda$ and an elliptic Andronov-Hopf point at $y \approx +\sqrt\lambda$, both of the generic type discussed in Chap. 5. See Fig. 9.3. They collide at the degenerate Andronov-Hopf point at the origin, for $\lambda = 0$. Finally, for $\lambda < 0$, no bifurcation occurs, i.e. the y-axis is normally stable and consists of stable foci.*

Again we expect the separatrices to split as discussed for the elliptic Andronov-Hopf point. In particular, the unstable set of the hyperbolic Andronov-Hopf point connects to a interval of the stable branch of the line of equilibria "behind" the elliptic Andronov-Hopf point. By Neishtadt averaging, for analytic vector fields, the size of the splitting is exponentially small in the distance of the two Andronov-Hopf points, that is in λ.

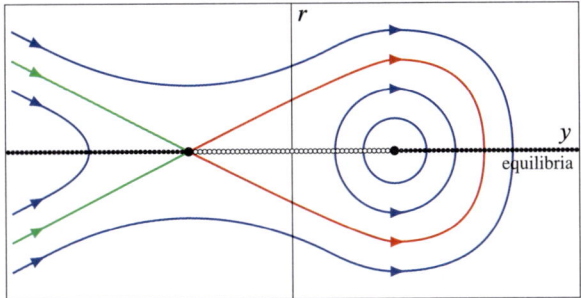

Fig. 9.3 Fold of Andronov-Hopf points. Normal form flow near a fold of Andronov-Hopf points alias fold of pitchfork bifurcations, $\lambda > 0$; stable set of the hyperbolic Andronov-Hopf point in *green*, unstable set in *red*

9.3 Planes of Equilibria

Along a plane of equilibria both singularities discussed before turn out to be equivalent, see Remark 9.4.

We start analogously to Sect. 8.3 with a line of Hopf points instead of transcritical points. Replacing the parameter λ discussed in Sect. 9.1 with an additional direction of a plane of equilibria, the drift along this manifold of equilibria is now a two-dimensional vector. Along generic one-dimensional curves, it will not vanish. Therefore, the drift singularity along curves of Hopf points is not characterized by a vanishing drift but rather by a drift direction tangential to the curve of Hopf points. (A drift in λ-direction was not possible in Sects. 9.1, 9.2 above.)

We consider a system

$$\begin{pmatrix} \dot{x} \\ \dot{y} \end{pmatrix} = F(x, y) = \begin{pmatrix} f(x, y) \\ g(x, y) \end{pmatrix}, \qquad x, y \in \mathbb{R}^2, \tag{9.19}$$

$x = (x_1, x_2)$, $y = (y_1, y_2)$, $f = (f_1, f_2)$, $g = (g_1, g_2)$, with the following properties:

(i) The y-plane consists of equilibria, $F(0, y) \equiv 0$.
(ii) There is a Poincaré-Andronov-Hopf point at the origin, i.e. the y-plane loses normal hyperbolicity at this point caused by a pair of purely imaginary eigenvalues of the linearization,

$$\partial_x f(0, 0) = \begin{pmatrix} 0 & -1 \\ 1 & 0 \end{pmatrix}.$$

(iii) This eigenvalue pair crosses the imaginary axis transversely, $\nabla_y \text{div}_x f(0, 0) \neq 0$. Without loss of generality, the gradient points in y_1-direction, we can choose

$\partial_{y_1} \mathrm{div}_x f(0,0) > 0$ and $\partial_{y_2} \mathrm{div}_x f(0,0) = 0$. By the implicit-function theorem, this gives rise to a curve of Hopf points tangential to the y_2-axis.

(iv) At the origin, the drift non-degeneracy transverse to the curve of Hopf points fails, $\Delta_x g_1(0,0) = 0$, with $\Delta = \partial_{x_1}^2 + \partial_{x_2}^2$.

(v) This drift degeneracy is transverse, i.e. the drift direction crosses the tangent to the curve of Hopf points with nonvanishing speed along the curve of Hopf points, $\partial_{y_1} \mathrm{div}_x f(0,0)\, \partial_{y_2} \Delta_x g_1(0,0) + \partial_{y_2}^2 \mathrm{div}_x f(0,0)\, \Delta_x g_2(0,0) > 0$.

(vi) The drift does not vanish at the origin, i.e. there is a component tangential to the curve of Hopf points, $\Delta_x g_2(0,0) < 0$.

Again, conditions (i–v) correspond to those of Sect. 9.1, the singularity described by (ii,iv) is robust under perturbations satisfying (i), provided the non-degeneracy conditions (iii,v,vi) hold. Signs in (iii,v,vi) are chosen without loss of generality, by switching signs of time, y_1, and y_2, if necessary.

The non-degeneracy condition (vi) indeed yields

$$\frac{\mathrm{d}}{\mathrm{d}y_2} \langle \nabla_y \mathrm{div}_x f, \Delta_x g \rangle (0, \vartheta(y_2), y_2) \Big|_{y_2 = 0} > 0, \qquad (9.20)$$

where $(x, y_1, y_2) = (0, \vartheta(y_2), y_2)$, $\vartheta(0) = 0$, $\vartheta'(0) = 0$, is the curve γ of Hopf points. Locally, we could reparametrize y to achieve $\vartheta \equiv 0$. Conditions (iii,v) then read:

$$\mathrm{div}_x f(0,0,y_2) \equiv 0, \qquad \partial_{y_1} \mathrm{div}_x f(0,0,0) > 0, \qquad \partial_{y_2} \Delta_x g_1(0,0,0) > 0. \qquad (9.21)$$

We assume this to hold.

In analogy to the previous sections, we find a normal form with additional rotationally symmetry,

$$\begin{aligned}
\dot{r} &= r h_r(r^2, y_1, y_2) + \text{h.o.t.}, \\
\dot{\varphi} &= 1, \\
\dot{y}_1 &= r^2 h_{y_1}(r^2, y_1, y_2) + \text{h.o.t.}, \\
\dot{y}_2 &= r^2 h_{y_2}(r^2, y_1, y_2) + \text{h.o.t.},
\end{aligned} \qquad (9.22)$$

Conditions (9.21) and (iv,vi) and are equivalent to $h_r(0,0,y_2) \equiv 0$, $\partial_{y_1} h_r(0,0,0) > 0$, $h_{y_1}(0,0,0) = 0$, $\partial_{y_2} h_{y_1}(0,0,0) > 0$, $h_{y_2}(0,0,0) < 0$.

We set $\tilde{y}_1 = h_r(0, y_1, y_2)$, to obtain

$$\begin{aligned}
\dot{r} &= r y_1 + r^3 \tilde{h}_r(r^2, \tilde{y}_1, y_2) + \text{h.o.t.}, \\
\dot{\varphi} &= 1, \\
\dot{\tilde{y}}_1 &= r^2 \tilde{h}_{y_1}(r^2, \tilde{y}_1, y_2) + \text{h.o.t.}, \\
\dot{y}_2 &= r^2 \tilde{h}_{y_2}(r^2, \tilde{y}_1, y_2) + \text{h.o.t.},
\end{aligned} \qquad (9.23)$$

still with $\tilde{h}_{y_1}(0,0,0) = 0$, $\partial_{y_2}\tilde{h}_{y_1}(0,0,0) > 0$, $\tilde{h}_{y_2}(0,0,0) < 0$. We drop tildes to simplify the notation, and omit the trivial φ direction as well as terms beyond normal form order. We obtain the (truncated) normal form

$$
\begin{aligned}
\dot{r} &= ry_1 + r^3 h_r(r^2, y), \\
\dot{y}_1 &= c_1 r^2 y_2 + r^2 h_{y_1}(r^2, y), \\
\dot{y}_2 &= c_2 r^2 + r^2 h_{y_2}(r^2, y),
\end{aligned}
\tag{9.24}
$$

with $h_{y_1}(0,0,0) = 0$, $\partial_{y_2}\tilde{h}_{y_1}(0,0,0) = 0$, $\tilde{h}_{y_2}(0,0,0) = 0$. The constants can be normalized to $c_1 = 1$ and $c_2 = -1$.

We rescale this system by

$$
\begin{aligned}
r &= \sigma^3 \tilde{r}, \\
y &= \sigma^4 \tilde{y}, \\
\lambda &= \sigma^2 \tilde{\lambda}, \\
t &= \sigma^{-4} \tilde{t}.
\end{aligned}
\tag{9.25}
$$

For $0 < \sigma \ll 1$, to leading order in σ we obtain the rescaled, truncated normal form

$$
\begin{aligned}
\dot{r} &= ry_1 + r\mathcal{O}(\sigma), \\
\dot{y}_1 &= r^2 y_2 + r^2 \mathcal{O}(\sigma), \\
\dot{y}_2 &= -r^2 + r^2 \mathcal{O}(\sigma).
\end{aligned}
\tag{9.26}
$$

Note that the rescaling used in Sect. 9.1 is not applicable, as it would be singular in the y_2-component. It is, however, reminiscent of the scaling used in Sect. 9.2.

System (9.26) is restricted to $r \geq 0$. Note the equilibrium plane $\{r = 0\}$ which is normally hyperbolic for $y_1 \neq 0$. The flow in y can be multiplied by $1/r^2 > 0$, keeping orbits and flow directions

$$
\begin{aligned}
y_1' &= y_2, \\
y_2' &= -1.
\end{aligned}
\tag{9.27}
$$

Solutions are just parabolas. It remains to discuss the convergence of r to zero along these curves. In (9.26) r can converge to 0 in forward time only for $y_1 < 0$ and in backward time only for $y_2 > 0$. Note also the elliptic Hopf points $y_1 = 0, y_2 < 0$ and the hyperbolic Hopf points $y_1 = 0, y_2 > 0$. We find a phase portrait as shown in Fig. 9.4.

Theorem 9.3 *Under conditions (i–vi), a truncated normal form of the vector field (9.19) is given by (9.26).*

For the normal form, a local neighborhood U of the origin is filled with

(i) *heteroclinic orbits connecting equilibria $\{y_1 > 0\}$ to equilibria $\{-\frac{3}{8}y_2^2 < y_1 < 0; \ y_2 < 0\}$, forming en extended "elliptic bubble" around the elliptic Hopf points $\{y_1 = 0; \ y_2 < 0\}$*

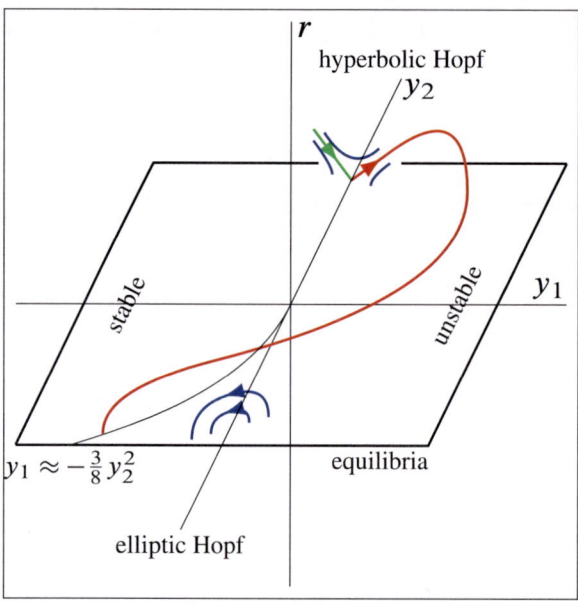

Fig. 9.4 Degenerate Hopf point on a plane of equilibria. Note the elliptic Hopf point, for $y_2 < 0$, and the hyperbolic Hopf point, for $y_2 > 0$, and the return of the unstable manifold of the half line of hyperbolic Hopf points to the plane of equilibria. Signs of y_1, y_2 are chosen w.l.o.g., the time-reversed phase portrait is also possible

(ii) orbits entering the neighborhood and converging to equilibria $\{y_1 < -y_2^2; \ y_2 < 0\} \cup \{y_1 < 0; \ y_2 \geq 0\}$ in forward time

The curve $\{y_1 = -\frac{3}{8}y_2^2; \ y_2 < 0\}$ is the limit of the strong unstable manifold of the hyperbolic Hopf points $\{y_1 = 0; \ y_2 > 0\}$ in the equilibrium plane. See Fig. 9.4

*   *The phase portrait of the full system has the same structure, although separatrix splitting occurs, i.e. the boundary of the two regions on the equilibrium plane depends on the phase angle. In particular, the hyperbolic Hopf points $\{y_1 = 0; \ y_2 > 0\}$ connect to a small wedge shaped region $\{y_1 \approx -\frac{3}{8}y_2^2; \ y_2 < 0\}$ on the equilibrium plane.*

Proof Equation (9.27) and normal stability of the half plane $\{y_1 < 0\}$ of equilibria ensures that all trajectories must converge to the plane of equilibria, in forward time. The domain of strong unstable manifolds of the normally unstable equilibria $\{y_1 > 0\}$ is bounded by the strong unstable manifolds of the hyperbolic Hopf points $\{y_1 = 0; \ y_2 > 0\}$. This hold true for their limit points on the y-plane.

Due to normal stability/instability of the half planes $\{y_1 \neq 0\}$ this remains true under perturbations.

It only remains to calculate the limit of the strong unstable manifold of the hyperbolic Hopf points $\{y_1 = 0; \ y_2 > 0\}$ on the equilibrium plane in the normal form (9.26). We multiply by $1/r$, keeping orbits, and drop higher-order terms.

$$\dot{r} = y_1,$$
$$\dot{y}_1 = ry_2, \tag{9.28}$$
$$\dot{y}_2 = -r.$$

Now consider a trajectory $(r(t), y_1(t), y_2(t))$ connecting a hyperbolic Hopf point $(0, 0, y_2^- > 0)$, for $t \to -\infty$, to a point on the y-plane $(r, y_1, y_2) = (0, y_1^+, y_2^+ < 0)$, for $t = 0$. We have $dy_1/dy_2 = -y_2$, see also (9.27) Thus $2y_1(t) = (y_2^-)^2 - y_2(t)^2$ and

$$\dot{r}(t) = \tfrac{1}{2}(y_2^-)^2 - \tfrac{1}{2}y_2(t)^2,$$
$$\dot{y}_2(t) = -r. \tag{9.29}$$

Separation of variables yields

$$\begin{aligned} 0 &= \int_{y_2^-}^{y_2^+} \left(\tfrac{1}{2}(y_2^-)^2 - \tfrac{1}{2}y_2(t)^2 \right) dy_2 \\ &= \tfrac{1}{6}(y_2^+)^3 - \tfrac{1}{2}(y_2^-)^2 y_2^+ + \tfrac{1}{3}(y_2^-)^3 \\ &= \tfrac{1}{6}(y_2^+ - y_2^-)^2 (y_2^+ + 2y_2^-) \end{aligned} \tag{9.30}$$

and therefore $y_2^+ = -2y_2^-$ and $y_1^+ = \tfrac{1}{2}((y_2^-)^2 - (y_2^+)^2) = -\tfrac{3}{8}(y_2^+)^2.$ □

Remark 9.4 There is no difference between drift and fold singularities, as discussed in Sects. 9.1 and 9.2 of one-parameter families of lines of equilibria, in the case of a plane of equilibria without parameters. Indeed, by a coordinate transformation of y, alone, the y_2-axis can be mapped to a parabola tangential to the y_1 axis.

Remark 9.5 Ignoring φ-dependence, i.e. without splitting of separatrices, the analysis in this chapter coincides with the analysis of the $\mathbb{Z}_2$ equivariant transcritical points with additional drift or fold degeneracy.

Chapter 10
Bogdanov-Takens Bifurcation

Classical Bogdanov-Takens bifurcation [13–16, 69, 70] is the most prominent bifurcation of codimension two. It is characterized by a versal unfolding of a nilpotent linearization with algebraically double and geometrically simple eigenvalue zero by two parameters. The rescaled normal form

$$\dot{x}_1 = \lambda_1 + \lambda_2 x_2 + x_2^2 \pm x_1 x_2,$$
$$\dot{x}_2 = x_1 \tag{10.1}$$

features a Hamiltonian core to leading order in the scaling

$$\begin{aligned}
x_1 &= \sigma^3 \tilde{x}_1, \\
x_2 &= \sigma^2 \tilde{x}_2, \\
\lambda_1 &= \sigma^4 \tilde{\lambda}_1, \\
\lambda_2 &= \sigma^2 \tilde{\lambda}_2, \\
t &= \sigma^{-1} \tilde{t}.
\end{aligned} \tag{10.2}$$

This integrable system is then perturbed by terms of higher order and facilitates expansions of curves in the parameter space of (classical) Poincaré-Andronov-Hopf bifurcations, saddle-node bifurcations and homoclinic orbits, see also [7, 37].

In analogy to this classical case, we call a singularity with an algebraically double and geometrically simple eigenvalue zero in the transverse directions to a surface—or plane—of equilibria a Bogdanov-Takens point. Alternatively, a one-parameter family of lines of equilibria could be considered. In Chaps. 8, 9, these settings differed. Here, it turns out that, for Bogdanov-Takens points, both settings lead to similar results, see (10.14).

Without parameters, we consider a system

$$\dot{z} = \begin{pmatrix} \dot{x} \\ \dot{y} \end{pmatrix} = F(z) = \begin{pmatrix} f(x, y) \\ g(x, y) \end{pmatrix}, \qquad x, y \in \mathbb{R}^2, \tag{10.3}$$

© Springer International Publishing Switzerland 2015
S. Liebscher, *Bifurcation without Parameters*, Lecture Notes in Mathematics 2117,
DOI 10.1007/978-3-319-10777-6_10

$x = (x_1, x_2)$, $y = (y_1, y_2)$, $f = (f_1, f_2)$, $g = (g_1, g_2)$, with the following properties:

(i) The y-plane consists of equilibria, $F(0, y) \equiv 0$.
(ii) At the origin, the linearization exhibits a nilpotent Jordan block,

$$DF(0) = \begin{pmatrix} 0 & 0 & 0 & 0 \\ 1 & 0 & 0 & 0 \\ 0 & 1 & 0 & 0 \\ 0 & 0 & 0 & 0 \end{pmatrix}. \tag{10.4}$$

(iii) This nilpotent linearization is versally unfolded by y. Specific non-degeneracy conditions (10.13) are given below. They yield a coordinate transformation such that

$$D_x f(0, 0) = \begin{pmatrix} -y_1 + y_2 & -y_1 \\ 1 & 0 \end{pmatrix}.$$

10.1   Normal Form

Note that (10.4) is the generic linearization for a geometrically simple and algebraically double eigenvalue in x-direction, i.e. for nilpotent $D_x f(0, 0)$ with one-dimensional kernel,

$$Df(0, 0) = \begin{pmatrix} 0 & 0 \\ 1 & 0 \end{pmatrix}. \tag{10.5}$$

Indeed, the y-plane of equilibria implies that

$$\text{range } DF(0) \cap \{x = 0\} \tag{10.6}$$

is invariant under $DF(0)$. Furthermore the kernel of $DF(0)$ has dimension at least 2, due to the manifold of equilibria, and generically no additional kernel vectors arise. Therefore the range of $DF(0)$ has also dimension 2. Due to (10.5), x_2 is in the range of $DF(0)$. Thus (10.6) is one-dimensional, generically, and w.l.o.g. orthogonal to the y_2-axis. This yields a linearization

$$DF(0) = \begin{pmatrix} 0 & 0 & 0 & 0 \\ 1 & 0 & 0 & 0 \\ c & 1 & 0 & 0 \\ 0 & 0 & 0 & 0 \end{pmatrix}, \tag{10.7}$$

and we are almost done. The shear in x given by $\tilde{x}_2 = cx_1 + x_2$ yields (10.4).

In [27] a normal form, adjusted to preserve the equilibrium manifold, has been calculated, see also Sect. 2.3. After suitable rescalings, the normal form can be written as the third-order equation

$$\ddot{v} + \dot{v}v = \varepsilon \left(\dot{v}(\lambda - v) + b\dot{v}^2 \right) + \mathcal{O}(\varepsilon^2), \tag{10.8}$$

with fixed parameters b, λ and ε. The (v, λ)-plane is the original y plane of equilibria, b depends on the nonlinearity, ε is a rescaling (or blow-up) parameter. Note the algebraically triple zero eigenvalue, double in the transverse directions $x = (\dot{v}, \ddot{v})$, for $\lambda = v = 0$.

In fact, a complete normal form procedure is not necessary. Start with the system (10.3) satisfying conditions (i,ii). Then the coordinate transformation

$$\begin{aligned}
\tilde{x}_1 &= Dg(x, y) \cdot F(x, y) = x_1 + \cdots, \\
\tilde{x}_2 &= g_1(x, y) &= x_2 + \cdots, \\
\tilde{y} &= y
\end{aligned} \tag{10.9}$$

yields the transformed system

$$\begin{aligned}
\dot{x}_1 &= 0 &+ h_1(x, y), \\
\dot{x}_2 &= x_1, \\
\dot{y}_1 &= x_2, \\
\dot{y}_1 &= 0 &+ h_4(x, y),
\end{aligned} \tag{10.10}$$

where we have dropped tildes to simplify the notation, expansions of h_1, h_4 start with quadratic terms, and vanish at the y-plane. We expand

$$h_1(x, y) = c_{11}x_1y_1 + c_{12}x_1y_2 + c_{21}x_2y_1 + c_{22}x_2y_2 + c_3x_2^2 + c_4x_1^2 + \mathcal{O}(|z|^3). \tag{10.11}$$

Then the linear transformation

$$\begin{aligned}
\tilde{x}_1 &= -c_{21}x_1, \\
\tilde{x}_2 &= -c_{21}x_2, \\
\tilde{y}_1 &= -c_{21}y_1 - c_{22}y_2, \\
\tilde{y}_2 &= (c_{12}c_{21}/c_{11} - c_{22})y_2,
\end{aligned} \tag{10.12}$$

with the non-degeneracy conditions

$$0 \neq c_{11}, c_{21}, c_{12}c_{21} - c_{11}c_{22} \tag{10.13}$$

yields

$$
\begin{aligned}
\dot{x}_1 &= a(-y_1 + y_2)x_1 - y_1x_2 + \hat{c}_3x_2^2 + \hat{c}_4x_1^2 + \mathcal{O}(|z|^3), \\
\dot{x}_2 &= x_1, \\
\dot{y}_1 &= x_2 + \mathcal{O}(|z|^2), \\
\dot{y}_2 &= \mathcal{O}(|z|^2),
\end{aligned}
\tag{10.14}
$$

with $a = c_{11}/c_{21} \neq 0$ and $\hat{c}_3 = -c_3/c_{21}$. Again we have dropped tildes to simplify notation.

Note the unfolding

$$
\begin{pmatrix} a(-y_1 + y_2) & -y_1 \\ 1 & 0 \end{pmatrix}
$$

of the nilpotent Jordan block. The y_2-axis $\{y_1 = 0\}$ is a family of transcritical points, Chap. 4, the diagonal $\{y_1 = y_2 > 0\}$ is a family of Poincaré-Andronov-Hopf points, Chap. 5. Both families emerge from the Bogdanov-Takens point. Again this reminds of the emergence of families of saddle-node bifurcations and Poincaré-Andronov-Hopf bifurcations from a classical Bogdanov-Takens point.

The final rescaling

$$
\begin{aligned}
x_1 &= (\varepsilon/a)^4\tilde{x}_1, \\
x_2 &= (\varepsilon/a)^3\tilde{x}_2, \\
y_1 &= (\varepsilon/a)^2\tilde{y}_1, \\
y_2 &= (\varepsilon/a)^2\tilde{y}_2, \\
t &= (\varepsilon/a)^{-1}\tilde{t},
\end{aligned}
\tag{10.15}
$$

and dropping tildes yield

$$
\begin{aligned}
\dot{x}_1 &= -y_1x_2 + \varepsilon\left((-y_1 + y_2)x_1 + bx_2^2\right) + \mathcal{O}(\varepsilon^2), \\
\dot{x}_2 &= x_1, \\
\dot{y}_1 &= x_2 + \mathcal{O}(|\varepsilon|^2), \\
\dot{y}_2 &= \mathcal{O}(|\varepsilon|^2),
\end{aligned}
\tag{10.16}
$$

with $b = \hat{c}_2/a = -c_3/c_{11}$. This is (10.8) with $v = y_1$, $\lambda = y_2$.

Here we also note that, to leading order, $\lambda = y_2$ is a classical parameter. Thus the cases of a plane of equilibria and a y_2-family of lines of equilibria are equivalent to leading order. Perturbations $\mathcal{O}(\varepsilon^2)$ will only introduce a small drift in y_2. This drift will preserve qualitative results relying on transverse splittings of order $\mathcal{O}(\varepsilon)$.

Figure 10.1 shows for the three resulting cases relevant parameter regions. Arrows indicate heteroclinic connection between equilibria of the given two-dimensional manifold. These pictures will be discussed in more detail in Sects. 10.5 and 10.6.

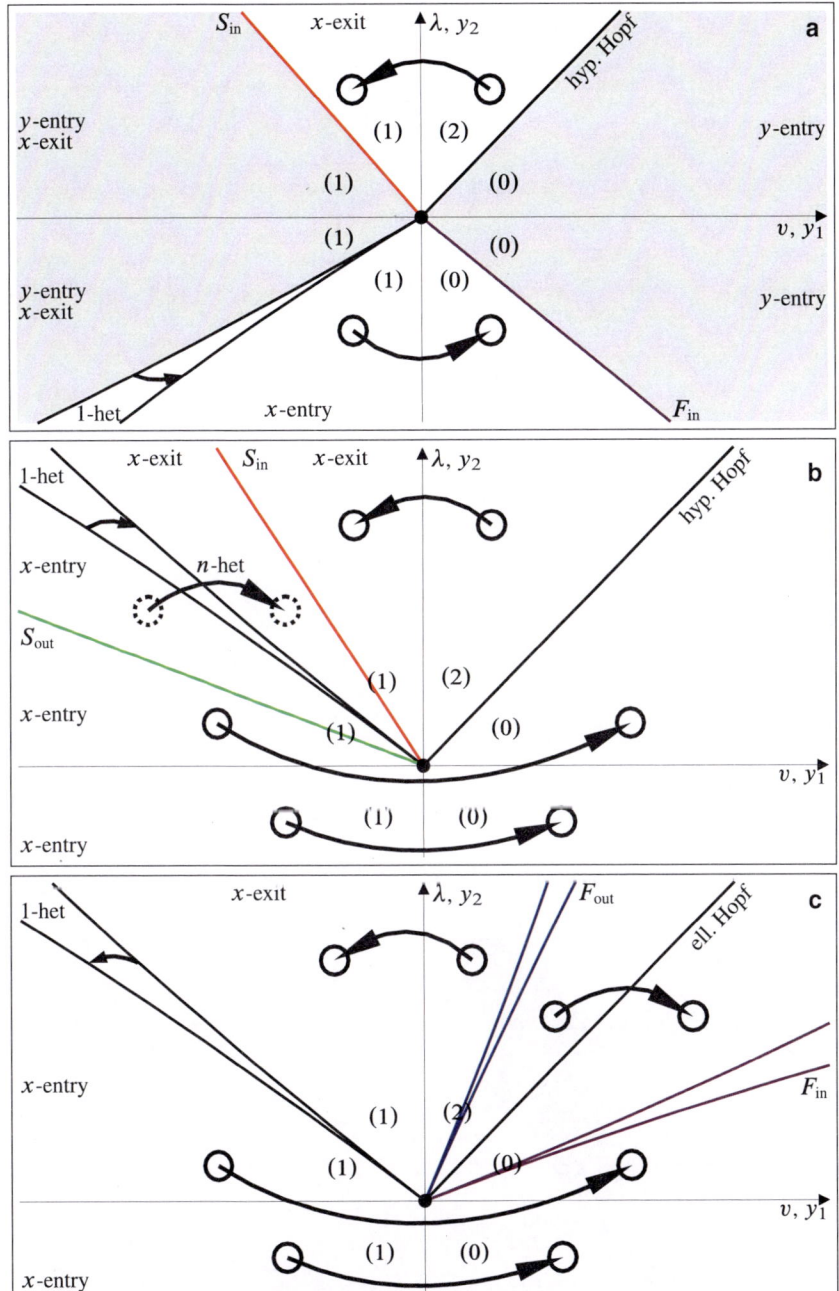

Fig. 10.1 Three cases of Bogdanov-Takens bifurcations without parameters. Unstable dimensions u of trivial equilibria $(0, y)$ of (10.8) are indicated by (u); "n-het" indicate saddle-saddle heteroclinics with n revolutions around the positive y_1-axis. Cases are distinguished by the coefficient b of (10.8): (**a**) $b < -17/12$, (**b**) $-17/12 < b < -1$, (**c**) $-1 < b$, see (10.53)

10.2 Integrable Core

For $\varepsilon = 0$, system (10.8) becomes completely integrable. This system represents the blow-up boundary, see Sect. 2.6. Two first integrals are then given by

$$
\begin{aligned}
\Theta &= \ddot{v} + \tfrac{1}{2}v^2, \\
H &= \tfrac{1}{2}\dot{v}^2 - \ddot{v}v - \tfrac{1}{3}v^3,
\end{aligned}
\tag{10.17}
$$

For fixed Θ, we obtain a Hamiltonian system, see Fig. 10.2. The phase space $(v, \dot{v}, \ddot{v}) = (y_1, x_2, x_1)$ can be parametrized by (v, Θ, H):

$$
\begin{aligned}
x_1 &= \Theta - \tfrac{1}{2}v^2 \\
x_2^2 &= -\tfrac{1}{12}q(v),
\end{aligned}
\tag{10.18}
$$

with Weierstrass polynomial

$$
q(v) = q(v; \Theta, H) = 4v^3 - 24\Theta v - 24H.
\tag{10.19}
$$

Note the scaling symmetry: for $\Theta > 0$,

$$
v^{\Theta, H}(t) = \Theta^{1/2} v^{1, \tilde{H}}(\Theta^{1/4} t)
\tag{10.20}
$$

is a solution of

$$
\ddot{v} + \tfrac{1}{2}v^2 - 1 = 0,
\tag{10.21}
$$

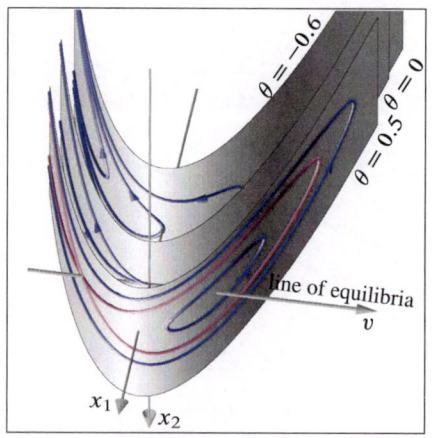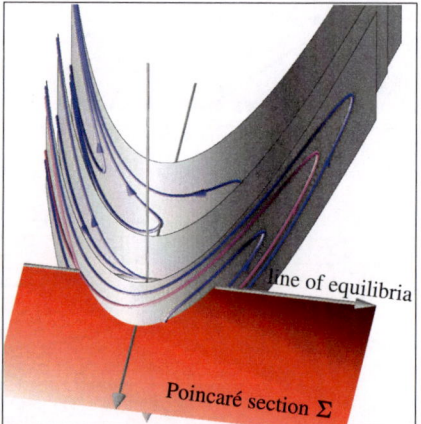

Fig. 10.2 Bogdanov-Takens point, integrable scaled flow, at order zero in ε

with energy

$$\tilde{H} = \Theta^{-3/2} H. \tag{10.22}$$

The region of bounded solutions of the integrable system is given by

$$|\tilde{H}| < \frac{2}{3}\sqrt{2}, \qquad \Theta > 0.$$

Simultaneously, this is a parametrization of the Poincaré section $\{x_2 = 0, x_1 < 0\}$, see Fig. 10.2. Indeed, the value of $y_1 = v$ on the Poincaré section Σ can be recovered from given (H, Θ) as the middle zero of the Weierstrass polynomial (10.19).

10.3 Poincaré Flow

For $\varepsilon > 0$, the quantities Θ, H are no longer conserved by (10.8). We find a slow drift

$$\begin{aligned}
\dot{\Theta} &= \varepsilon\left[(\Theta - \tfrac{1}{2}v^2)(-v + \lambda) - \tfrac{1}{12}bq(v)\right], \\
\dot{H} &= -\varepsilon y\left[(\Theta - \tfrac{1}{2}v^2)(-v + \lambda) - \tfrac{1}{12}bq(v)\right]
\end{aligned} \tag{10.23}$$

of order $\mathscr{O}(\varepsilon)$. We study this drift in (Θ, H) for the return map Π^ε to the Poincaré section $\{x_2 = 0, x_1 < 0\}$. To leading order, the drift is given by its integral over the periodic orbits of the integrable system. Therefore, it is the time-ε map of the Poincaré flow

$$\begin{pmatrix} \dot{\Theta} \\ \dot{H} \end{pmatrix} = \int_0^{T^0} \left[(\Theta - \tfrac{1}{2}v^2)(-v + \lambda) - \tfrac{1}{12}bq(v)\right] \begin{pmatrix} 1 \\ -v \end{pmatrix} dt \tag{10.24}$$

on the Poincaré section. The Poincaré return time T^0 is given by the minimal period of the periodic orbit $v(t)$ of the integrable order-zero vector field. Moreover, the flow (10.24) can be calculated in terms of Weierstrass elliptic integrals.

$$J_k(\Theta, H) = \int_0^{T^0} (v(t))^k\, dt = \Theta^{k/2 - 1/4} J_k(\tilde{H}). \tag{10.25}$$

Specifically, we find using (10.19), (10.22):

$$\begin{aligned}
\dot{\Theta} &= (\tfrac{1}{2} - \tfrac{b}{3})\Theta^{5/4} J_3(\tilde{H}) - \tfrac{\lambda}{2}\Theta^{3/4} J_2(\tilde{H}) \\
&\quad + (2b - 1)\Theta^{5/4} J_1(\tilde{H}) + (\lambda\Theta^{3/4} + 2b\Theta^{5/4}\tilde{H}) J_0(\tilde{H}), \\
\dot{H} &= -(\tfrac{1}{2} - \tfrac{b}{3})\Theta^{5/4} J_4(\tilde{H}) + \tfrac{\lambda}{2}\Theta^{3/4} J_3(\tilde{H}) \\
&\quad - (2b - 1)\Theta^{5/4} J_2(\tilde{H}) - (\lambda\Theta^{3/4} + 2b\Theta^{5/4}\tilde{H}) J_1(\tilde{H}), \\
\dot{\tilde{H}} &= -\tfrac{3}{2}\Theta^{-1}\tilde{H}\dot{\Theta} + \Theta^{-3/2}\dot{H} \\
&= \Theta^{-1}\dot{\Theta}\left(-\tfrac{3}{2}\tilde{H} + \Theta^{-1/2}\dot{H}/\dot{\Theta}\right).
\end{aligned} \tag{10.26}$$

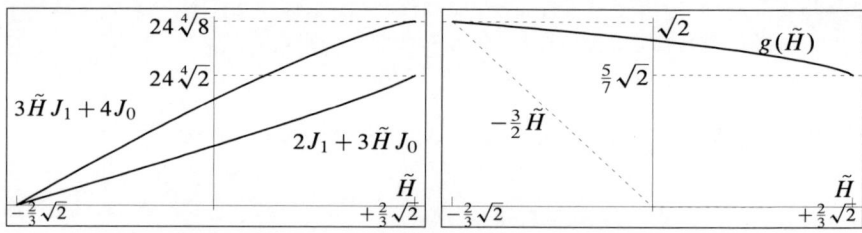

Fig. 10.3 Plots of the nonlinearities $2J_1 + 3\tilde{H}J_0$, $3\tilde{H}J_1 + 4J_0$, and $g(\tilde{H})$

In [27], the recursion relations

$$
\begin{aligned}
J_0 &= J_0(\tilde{H}), \\
J_1 &= J_1(\tilde{H}), \\
J_2 &= 2J_0, \\
J_3 &= \tfrac{6}{5}(3J_1 + 2\tilde{H}J_0), \\
J_4 &= \tfrac{12}{7}(2\tilde{H}J_1 + 5J_0)
\end{aligned}
\tag{10.27}
$$

have been used to calculate the integrals and the Poincaré flow

$$
\begin{aligned}
\dot{\Theta} &= \tfrac{2}{5}\Theta^{5/4}(b+1)(2J_1 + 3\tilde{H}J_0), \\
\dot{\tilde{H}} &= \tfrac{2}{5}\Theta^{1/4}(b+1)(2J_1 + 3\tilde{H}J_0)\left(\tfrac{\lambda}{b+1}\Theta^{-1/2} - \tfrac{3}{2}\tilde{H} - \alpha g(\tilde{H})\right),
\end{aligned}
\tag{10.28}
$$

with

$$
\begin{aligned}
\alpha &= \frac{b+2}{b+1} = 1 + \frac{1}{b+1}, \\
g(\tilde{H}) &= \frac{5}{7}\,\frac{3\tilde{H}J_1(\tilde{H}) + 4J_0(\tilde{H})}{2J_1(\tilde{H}) + 3\tilde{H}J_0(\tilde{H})},
\end{aligned}
\tag{10.29}
$$

where $b \neq -1$ is assumed as a genericity requirement. However, numerical observations were necessary to find g to be monotone in $\tilde{H}$ between the analytically calculated boundary values $\sqrt{2}$ and $5\sqrt{2}/7$, see Fig. 10.3. We close this gap in the following section and thank Stephan van Gils for pointing out this approach.

10.4 Elliptic Integrals and the Ricatti Equation

We recall the Hamiltonian structure

$$
\frac{1}{2}\dot{v}^2 = -\frac{1}{24}q(v;\tilde{H}) = -\frac{1}{6}v^3 + v + \tilde{H}
\tag{10.30}
$$

and the elliptic integrals

$$J_k(\tilde{H}) = \int_0^T v(t)^k \, dt = 2 \int_{e_2}^{e_1} \frac{v^k}{\dot{v}} \, dv, \tag{10.31}$$

where $e_1 > e_2 > e_3$ denote the three real zeros of q.
   Now, we define

$$I_k(\tilde{H}) = \int_0^T \dot{v}(t)^2 v(t)^k \, dt = 2 \int_{e_2}^{e_1} \dot{v} v^k \, dv. \tag{10.32}$$

Then, we can view $\dot{v}$ as a function of v and $\tilde{H}$, at least locally, and obtain from (10.30)

$$\dot{v} \frac{d}{d\tilde{H}} \dot{v} = 1,$$
$$\frac{d}{d\tilde{H}} I_k = J_k. \tag{10.33}$$

Differentiation of (10.30) by v,

$$\dot{v} \frac{d}{dv} \dot{v} + \frac{1}{2} v^2 - 1 = 0, \tag{10.34}$$

multiplication by $v^k/\dot{v}$ and integration over one period yields the recursion

$$- k I_{k-1} + \frac{1}{2} J_{k+2} - J_k = 0, \qquad k \geq 0. \tag{10.35}$$

For $k = 0, \dots, 3$ we obtain explicitly

$$\begin{aligned}
\tfrac{1}{2} J_2 &= J_0, \\
\tfrac{1}{2} J_3 &= J_1 + I_0, \\
\tfrac{1}{2} J_4 &= J_2 + 2I_1 = 2J_0 + 2I_1, \\
\tfrac{1}{2} J_5 &= J_3 + 3I_2 = 2J_1 + 3I_2 + 2I_0.
\end{aligned} \tag{10.36}$$

A second recursion formula results from the multiplication of $\tilde{H}$ and J_k:

$$- \tilde{H} J_k + \tfrac{1}{2} I_k + \tfrac{1}{6} J_{k+3} - J_{k+1} = 0. \tag{10.37}$$

Together with (10.35), we can obtain the recursion relations (10.27) of J_k alone and thus have calculated the Poincaré flow (10.28). In addition, we find the expressions

$$\tilde{H} J_0 = \tfrac{5}{6} I_0 - \tfrac{2}{3} J_1,$$
$$\tilde{H} J_1 = \tfrac{7}{6} I_1 - \tfrac{4}{3} J_0. \tag{10.38}$$

Direct computation of the integrals (10.32) in the boundary cases $\tilde{H} = -\tfrac{2}{3}\sqrt{2}$, $e_1 = e_2 = \sqrt{2}$ and $\tilde{H} = \tfrac{2}{3}\sqrt{2}$, $e_1 = 2\sqrt{2}$, $e_2 = e_3 = -\sqrt{2}$ then yields

$$2J_1 + 3\tilde{H} J_0 = 5I_0 = \begin{cases} 0 & \text{at } \tilde{H} = -\tfrac{2}{3}\sqrt{2} \\ 2^{1/4} \cdot 24 & \text{at } \tilde{H} = \tfrac{2}{3}\sqrt{2} \end{cases}$$

$$3J_1 \tilde{H} + 4J_0 = 7I_1 = \begin{cases} 0 & \text{at } \tilde{H} = -\tfrac{2}{3}\sqrt{2} \\ 2^{3/4} \cdot 24 & \text{at } \tilde{H} = \tfrac{2}{3}\sqrt{2} \end{cases} \tag{10.39}$$

and will be used later to discuss the Poincaré flow near the line of equilibria.

System (10.38) can be rewritten as

$$\frac{d}{d\tilde{H}} \begin{pmatrix} I_0 \\ I_1 \end{pmatrix} = \begin{pmatrix} J_0 \\ J_1 \end{pmatrix} = \frac{1}{\tfrac{8}{9} - \tilde{H}^2} \begin{pmatrix} -\tfrac{5}{6}\tilde{H} & \tfrac{7}{9} \\ \tfrac{10}{9} & -\tfrac{7}{6}\tilde{H} \end{pmatrix} \begin{pmatrix} I_0 \\ I_1 \end{pmatrix}. \tag{10.40}$$

Note that the first factor is positive for $|\tilde{H}| < \tfrac{2}{3}\sqrt{2}$.

The quotient $g(\tilde{H})$, see (10.29), turns out to be given by

$$g(\tilde{H}) = \frac{I_1}{I_0}. \tag{10.41}$$

The integral I_0 is positive, for $\tilde{H} > -\tfrac{2}{3}\sqrt{2}$, by definition. Positiveness of I_1 is proved by the following argument:

At the center equilibrium, $\tilde{H} = -\tfrac{2}{3}\sqrt{2}$, the value of I_1 vanishes. For slightly larger values of $\tilde{H}$, the corresponding periodic orbit is near the center equilibrium at $v = \sqrt{2} > 0$ and, therefore, I_1 has to be positive. Now assume that at some point, $-\tfrac{2}{3}\sqrt{2} < \tilde{H}^* < \tfrac{2}{3}\sqrt{2}$, there exists a zero, $I_1(\tilde{H}^*) = 0$. Chose $\tilde{H}^*$ to be minimal. Then the second component of (10.40) reads

$$\left. \frac{d}{d\tilde{H}} I_1 \right|_{\tilde{H}=\tilde{H}^*} = \frac{1}{\tfrac{8}{9} - \tilde{H}^2} \frac{10}{9} I_0 > 0. \tag{10.42}$$

This is a contradiction to the positiveness of I_1 on the left boundary. We have proved the positiveness of I_1:

$$\left. \begin{matrix} I_0 > 0 \\ I_1 > 0 \end{matrix} \right\} \quad \text{for} \quad -\tfrac{2}{3}\sqrt{2} < \tilde{H} < \tfrac{2}{3}\sqrt{2}. \tag{10.43}$$

The last open statement claims the monotonicity of the quotient (10.41), see also (10.29). The proof uses the Ricatti equation which usually is written for the inverse quotient I_0/I_1. Again, we differentiate by $\tilde{H}$ and use (10.40) to obtain the Ricatti equation:

$$
\frac{\mathrm{d}}{\mathrm{d}\tilde{H}} g(\tilde{H}) = \frac{J_1}{I_0} - \frac{I_1 J_0}{I_0^2}
$$
$$
= -\frac{1}{\frac{8}{9} - \tilde{H}^2} \left(\frac{7}{9} g^2 + \frac{1}{3} \tilde{H} g - \frac{10}{9} \right).
\tag{10.44}
$$

We determine the value of g, in fact of its continuation, at the center equilibrium $v = \sqrt{2}$ by L'Hôpital's rule, see (10.32, 10.41):

$$
g(-\tfrac{2}{3}\sqrt{2}) = \sqrt{2}.
\tag{10.45}
$$

The other boundary value is given by (10.39):

$$
g(\tfrac{2}{3}\sqrt{2}) = \tfrac{5}{7}\sqrt{2}.
\tag{10.46}
$$

Assume that, contrary to the monotonicity claim, there exists a local extremum of g at some point $-\frac{2}{3}\sqrt{2} < \tilde{H}^* < \frac{2}{3}\sqrt{2}$. Then we obtain from (10.44) at the local extremum:

$$
\frac{\mathrm{d}}{\mathrm{d}\tilde{H}} g(\tilde{H}^*) = 0,
$$
$$
\frac{\mathrm{d}^2}{\mathrm{d}\tilde{H}^2} g(\tilde{H}^*) = -\frac{1}{\frac{8}{9} - (\tilde{H}^*)^2} \frac{g(\tilde{H}^*)}{3}.
\tag{10.47}
$$

We already know that g is positive in the considered domain. Therefore, only local maxima of g are possible but no local minima. However, the existence of a local maximum of g at $\tilde{H}^*$ without an accompanying minimum would require that $g(\tilde{H}^*)$ is larger than the value at the left boundary, $g(\tilde{H}^*) > \sqrt{2}$. The contradiction is again shown by the Ricatti equation (10.44):

$$
0 = \frac{\mathrm{d}}{\mathrm{d}\tilde{H}} g(\tilde{H}^*)
$$
$$
= -\frac{1}{\frac{8}{9} - (\tilde{H}^*)^2} \left(\frac{7}{9} g^2 + \frac{1}{3} \tilde{H}^* g - \frac{10}{9} \right).
$$
$$
> -\frac{1}{\frac{8}{9} - (\tilde{H}^*)^2} \left(\left(\frac{7}{9}\sqrt{2} - \frac{2}{9}\sqrt{2} \right) g - \frac{10}{9} \right).
\tag{10.48}
$$
$$
> -\frac{1}{\frac{8}{9} - (\tilde{H}^*)^2} \left(\frac{5}{9} 2 - \frac{10}{9} \right).
$$
$$
= 0.
$$

Thus, finally, the

$$\text{monotone decay of } g(\tilde{H}) \text{ from } \sqrt{2} \text{ to } \tfrac{5}{7}\sqrt{2} \qquad (10.49)$$

for $\tilde{H}$ going from $-\tfrac{2}{3}\sqrt{2}$ to $\tfrac{2}{3}\sqrt{2}$, as shown in Fig. 10.3, is proved. A straightforward calculation using (10.44) and (10.49) also shows that

$$\frac{d}{d\tilde{H}} \frac{g(\tilde{H})}{\tilde{H}} < 0, \qquad (10.50)$$

for $0 < |\tilde{H}| < \tfrac{2}{3}\sqrt{2}$. With the boundary value (10.45) this yields in particular

$$g(\tilde{H}) > -\tfrac{3}{2}\tilde{H}, \qquad \text{for} \quad -\tfrac{2}{3}\sqrt{2} < \tilde{H} < \tfrac{2}{3}\sqrt{2}, \qquad (10.51)$$

which is also depicted in Fig. 10.3.

10.5 Discussion of the Poincaré Flow

We now return to the Poincaré flow (10.28), and note again that we assumed $b \neq -1$. Moreover, $\Theta > 0$ is invariant, and $2J_1 + 3\tilde{H}J_0 > 0$ except for the centers $\tilde{H} = -\tfrac{2}{3}\sqrt{2}$. We parametrize $(\Theta, \tilde{H})$-orbits over $\tau = \ln\Theta - \ln(\tfrac{\lambda}{b+1})^2$ and write $' = \tfrac{d}{d\tau}$. This simplifies (10.28) to

$$\tilde{H}'(\tau) = \pm e^{-\tau/2} - \frac{3}{2}\tilde{H} - \alpha g(\tilde{H}). \qquad (10.52)$$

The flow profiles of this equation are comparatively easy to discuss. Note that orbits of (10.52) and (10.28) coincide with reversed time direction for $b < -1$. The equilibrium v-axis, a cusp in (Θ, H) coordinates, transforms to the top (saddles) and bottom (foci) horizontal boundaries, with $v = 0$ shifted to $\tau = -\infty$. Since τ and $\tilde{H}$ are constants of the flow, for $\varepsilon = 0$, they represent slow drifts superposed on the unperturbed periodic motion, for small $\varepsilon > 0$ and $\tilde{H}$ between the top and bottom boundaries, $\tilde{H}^2 < \tfrac{8}{9}$. The top boundary also represents homoclinic orbits to the saddles, for $\varepsilon = 0$.

For each of the signs $\pm$, given by $\text{sign}(\lambda(b+1))$, three different regimes of the real parameter $\alpha = 1 + 1/(b+1)$ arise:

Case (A) $b < -17/12 \iff -7/5 < \alpha < 1$,

Case (B) $-17/12 < b < -1 \qquad \iff \qquad \alpha < -7/5$, $\qquad (10.53)$

Case (C) $-1 < b \qquad\qquad \iff \qquad 1 < \alpha$.

The phase portraits of the six resulting cases are shown in Fig. 10.4.

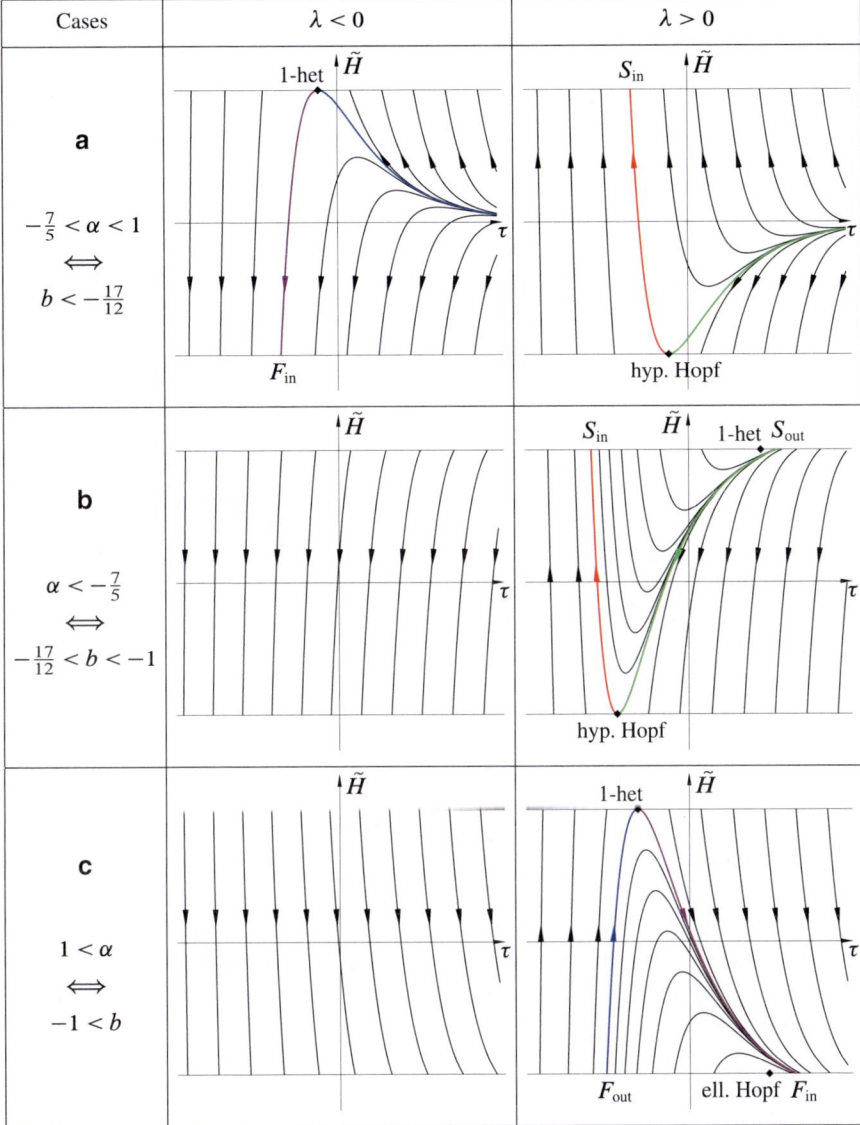

Fig. 10.4 Bogdanov-Takens point, Poincaré flow. The Poincaré flow given by the drift in the conserved quantities integrated over the periodic orbits of the integrable flow, Fig. 10.2. Coordinates are $\tau = \ln \Theta + \tau_0$, $\tilde{H}$, see (10.52), flow direction is according to (10.28). The Poincaré map of the full system amounts to a first-order discretization of this flow

In each picture, the bottom horizontal line $\tilde{H} = -\frac{2}{3}\sqrt{2}$ refers to the equilibrium half line $x_1 = \dot{v} = 0$, $x_2 = \ddot{v} = 0$, $y_1 = v = \sqrt{2\Theta} > 0$, for any $y_2 = \lambda$; in terms of the original Takens-Bogdanov system (10.3) or its normal form (10.8), see also Fig. 10.1. Consistently with this observation, both $\dot{\Theta}$ and $\dot{\tilde{H}}$ vanish along this line, because $2J_1 + 3\tilde{H}J_0 = 0$ at $\tilde{H} = -\frac{2}{3}\sqrt{2}$; see (10.39). In the original coordinates (x_1, x_2, y), these equilibria are normally hyperbolic, except along the Hopf line $\lambda = y > 0$ which manifests itself by a horizontal tangent $\tilde{H}' = 0$ at $\tilde{H} = -\frac{2}{3}\sqrt{2}$. The (strict) unstable dimension of the equilibria switches from 2, for $0 < y < \lambda$, to 0 for $0 < \lambda < y$; this is easily detected in the right half of Fig. 10.4.

For fixed $\lambda < 0$, in the left half of Fig. 10.4, we just as easily detect strict normal stability of the equilibria $x_1 = x_2 = 0$, $y > 0$.

For very negative τ, i.e. small $\Theta > 0$, we obtain almost vertical orbits which connect the horizontal boundaries $\tilde{H} = \pm\frac{2}{3}\sqrt{2}$ in very short "time" intervals τ. Indeed, $|\tilde{H}| \leq \frac{2}{3}\sqrt{2}$ and $\frac{5}{7}\sqrt{2} \leq g \leq \sqrt{2}$ are uniformly bounded in the region of interest, see Fig. 10.3 and (10.49). The direction, with proper reversal for $b < -1$, is easily determined.

For very positive τ, as well as for $\lambda = 0$, the exponential term disappears and we are left with the autonomous limiting equation

$$-\tilde{H}'(\tau) = \tfrac{3}{2}\tilde{H} + \alpha g(\tilde{H}). \tag{10.54}$$

Again using our bounds $\frac{5}{7}\sqrt{2} \leq g \leq \sqrt{2}$, $-\frac{3}{2}\tilde{H} \leq g$, see (10.49, 10.51), the right-hand side possesses a zero if, and only if,

$$-\tfrac{7}{5} < \alpha < 1. \tag{10.55}$$

The zero is unique due to the monotonicity of $g/\tilde{H}$ and g, see (10.50) and (10.49). Only in this case (A), therefore, the orbits of (10.52) do not connect the horizontal boundaries $\tilde{H} = \pm\frac{2}{3}\sqrt{2}$ within finite "time" τ, for all large positive τ, as they did for very negative τ and still do for large positive τ in the other cases (B), (C). Rather, in case (A), the unique zero of $\frac{3}{2}\tilde{H} + \alpha g(\tilde{H})$ provide semi-invariant regions, for large $\tau > 0$, which prevent orbits from connecting the horizontal boundaries. Instead,

$$\lim_{\tau \to +\infty} \tilde{H}(\tau) \in (-\tfrac{2}{3}\sqrt{2}, +\tfrac{2}{3}\sqrt{2}) \tag{10.56}$$

exists for all orbits starting at sufficiently large $\tau > 0$. This accounts for the shaded regions labeled "y-entry", in Fig. 10.1a. Note that the backwards escape time $t < 0$ is finite in terms of the original system (10.28), due to positivity of $2J_1 + 3\tilde{H}J_0 = \frac{5}{2}I_0$, see (10.38), (10.43).

The upper horizontal boundary $\tilde{H} = +\frac{2}{3}\sqrt{2}$ does not only indicate the half line of saddle equilibria $x_1 = x_2 = 0$, $y = -\sqrt{2\Theta} < 0$, which can be discussed analogously to the line $\tilde{H} = -\frac{2}{3}\sqrt{2}$ treated above. It also characterizes the fate the family of homoclinic orbits that exists for $\varepsilon = 0$: these homoclinic orbits break due to the drift for $\varepsilon > 0$. Indeed, if we insert the nonzero limiting values (10.39), (10.46) in (10.28), at $\tilde{H} = \frac{2}{3}\sqrt{2}$, then $\dot{\Theta} \neq 0$ along this line, and $\tilde{H}'$ vanishes only at the simple zero

$$\pm e^{-\tau/2} = \tfrac{3}{2}\tilde{H} + \alpha g(\tilde{H}) = \sqrt{2}\frac{12b + 17}{7(b + 1)} \tag{10.57}$$

of the right hand side of (10.52). In terms of (10.23) and the original (scaled) variables $y = -\sqrt{2\Theta} = \mp\sqrt{2}e^{\tau/2}\lambda/(b + 1)$ this occurs along the asymptote

$$\lambda = -\tfrac{1}{7}(12b + 17)y. \tag{10.58}$$

The corresponding points in Fig. 10.4, and lines in Fig. 10.1, are labeled "1-het". They correspond to zeros of an associated Melnikov function and to saddle-saddle heteroclinic orbits, in the original system. The zero is simple, therefore the splitting is transverse and thus robust under small perturbations. The remaining boundary regions $\tilde{H} = \frac{2}{3}\sqrt{2}$ where $\dot{\tilde{H}}$ is positive or negative, respectively, indicate a splitting of the homoclinic boundary of the periodic region, such that orbits escape in backward or forward time from a neighborhood U of the origin in the equilibrium plane $x_1 = x_2 = 0$, see also Fig. 10.5. In Fig. 10.1 this behavior is marked as "x-entry/exit".

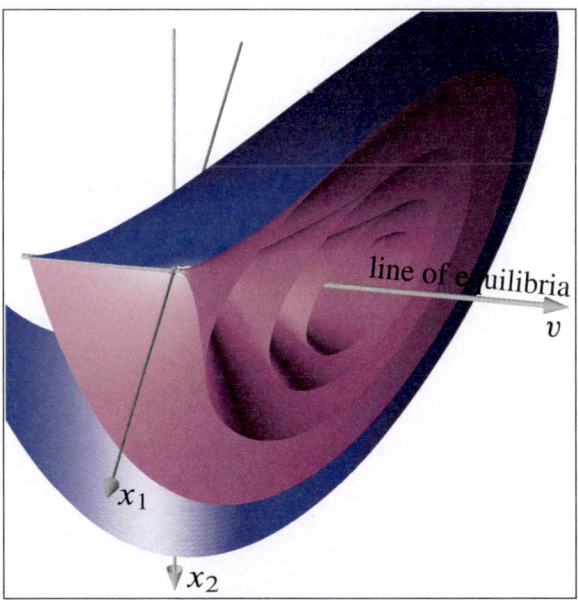

Fig. 10.5 Bogdanov-Takens point, splitting of manifolds. Splitting of the homoclinic boundary of the periodic region due to the slow drift of order $\mathcal{O}(\varepsilon)$ in (10.8) and (10.23), for Θ close to 0 and $\lambda < 0$. Compare with Figs. 10.2 and 10.4. Center-stable manifold of the saddle equilibria in *blue*, center-unstable manifolds in *magenta*

10.6   Poincaré Return Map and Bounded Solutions

To complete our discussion of the local dynamics near a Bogdanov-Takens point in the center manifold of the full system (10.3), we recall the three approximation steps that we have applied:

(a) truncation to second order normal form (10.14)
(b) omission of scaling terms of order ε^2 and higher (10.16)
(c) approximation of the Poincaré map Π^ε by the time-ε map of the Poincaré flow (10.24) which is the integral of the order-1 flow (10.23) over the periodic orbits of the integrable order-0 flow (Takens:eqConservedQuantities).

To ensure that our original variables $((\varepsilon/a)^4 x_1, (\varepsilon/a)^3 x_2, (\varepsilon/a)^2 y_1, (\varepsilon/a)^2 y_2)$ cover a neighborhood U of the origin in $\mathbb{R}^4$, we fix $C > 0$ arbitrarily large, consider the scaled variables (x_1, x_2, v, λ) in a ball of radius C, and analyze the complete, non-truncated rescaled dynamics for $0 \le \varepsilon < \varepsilon_0 = \varepsilon_0(C)$.

In terms of the variables $(\tau, \tilde{H}, v)$, $\tau = \ln \Theta + \tau_0$, $\tau_0 = -\ln(\frac{\lambda}{b+1})^2$, we immediately see that solutions that do not intersect the Poincaré section

$$\Sigma = \{(\tau, \tilde{H}, v) \mid \tau \in \mathbb{R}, \ |\tilde{H}| < \tfrac{2}{3}\sqrt{2}, \ v = e_2\} \tag{10.59}$$

become unbounded, or else belong to some equilibrium in the closure $\bar{\Sigma}$ or to its strong stable or strong unstable manifold $W^{ss}(v)$, $W^{uu}(v)$. As before $e_2 = e_2(\Theta, H)$ denotes the middle zero of the Weierstrass polynomial $q(v)$, where indeed $x_2 = 0$; see (10.18), (10.19), and Fig. 10.2.

We are interested in bounded nonstationary solutions which remain in U and thus intersect Σ. The Poincaré map Π^ε, wherever defined on Σ, is just some first-order discretization of the Poincaré flow (10.24) with time step ε. Indeed, to order 0 in ε, the system is integrable and $\Pi^0 \equiv$ id on Σ, see Sect. 10.2. To order 1 in ε, the system is given by the flow (10.23), and Π^ε is a first-order discretization of the Poincaré flow (c). Finally, all approximations (a), (b) perturb the Poincaré map Π^ε by terms of order ε^2 or higher order, therefore the map Π^ε remains a first-order discretization of the unperturbed Poincaré flow (10.24).

The Poincaré flow (10.24) is strongly related to the averaging approximation of the order-1 flow (10.23). Standard averaging, see for example [7], aims at eliminating the oscillations of v from the slow flow of (Θ, H) in (10.23). The elimination is achieved by successive v-dependent transformations of (Θ, H). These transformations successively lead to autonomous vector fields for (Θ, H), independent of v, with v-dependent corrections of order ε^N, $N = 2, 3, \ldots$. The first step, $N = 2$, replaces y^k in (10.23) by its time average

$$\frac{1}{T} \int_0^{T^0} (y(t))^k \, dt = \frac{J_k}{J_0},$$

over the unperturbed T-periodic solution $v(t)$, see (10.25). Replacing J_k by J_k/J_0, everywhere, converts our Poincaré flow (10.26) to the first averaged flow of the standard approach. The Poincaré flow and the averaged flow just differ by an Euler multiplier $T^0 = T^0(\Theta, H) = J_0$. Approximation of a vector field by its average relies on the separation of time scales of the fast averaged motion in v and the slow drift in (Θ, H). The separation holds as long as the return time T^0 remains finite.

At the homoclinic orbits we find $J_0 = \infty$, $\tilde{H} = \frac{2}{3}\sqrt{2}$, time scales do not separate anymore and the interpretation of the Poincaré flow as averaging is no longer valid. Instead, the Poincaré flow near $\tilde{H} = \frac{2}{3}\sqrt{2}$ can be interpreted directly in terms of Melnikov functions. In general, Melnikov functions are given as integrals of the perturbation of a vector field applied to nontrival solutions of the adjoint variational equation of the unperturbed vector field along a homoclinic orbit and describe the splitting of said orbit under the perturbation. See [21, 31] for a background.

But even without resorting to Melnikov theory, we see how the Poincaré flow (10.24) describes the splitting of the homoclinic boundary of the periodic region of the integrable core (10.17) near $\tilde{H} = \frac{2}{3}\sqrt{2}$. Indeed, as long as the solution of (10.23) starting at $(\Theta_0, H_0) \in \Sigma$ returns to the Poincaré section Σ after time T^ε, the return map is given by

$$\int_0^{T^\varepsilon} \left[(\Theta(t) - \tfrac{1}{2}v(t)^2)(-v(t) + \lambda) - \tfrac{1}{12}bq(v(t); \Theta(t), H(t)) \right] \begin{pmatrix} 1 \\ -v(t) \end{pmatrix} dt. \tag{10.60}$$

It converges to

$$\int_0^{T^0} \left[(\Theta_0 - \tfrac{1}{2}v(t)^2)(-v(t) + \lambda) - \tfrac{1}{12}bq(v(t); \Theta_0, H_0) \right] \begin{pmatrix} 1 \\ -v(t) \end{pmatrix} dt \tag{10.61}$$

of order ε, uniformly in the return time $T^\varepsilon \leq \infty$. For saddle equilibria $(\Theta_0, H_0) \in \partial\Sigma$ where $T^\varepsilon = +\infty$, we only have to replace the return point $(\Theta(T^\varepsilon), H(T^\varepsilon))$ in Σ by the return point of the strong unstable manifold of (Θ_0, H_0). An analogous argument holds for the reversed flow. In any case, we find (10.60) and (10.61) do differ by $\mathcal{O}(\varepsilon^2)$. Note that the second integral (10.61) defines the Poincaré flow (10.24).

In particular, at points of the line $\tilde{H} = \frac{2}{3}\sqrt{2}$ at which the Poincaré flow (10.24) points outwards, $H' > 0$, the strong unstable manifold of the corresponding saddle equilibrium does not return, whereas the strong stable manifold returns (under the backward flow). At points of the Poincaré flow points inwards, $H' < 0$, the strong unstable manifold of the corresponding saddle equilibrium returns, whereas the strong stable manifold does not return (under the backward flow). In the last section, we have already determined the simple zeros (10.57) of $\tilde{H}'$. They yield a transverse intersection of the family of strong unstable manifolds—alias center unstable manifold—with the upper boundary or with the family of strong stable manifolds. Transversality also holds for the full system which differs from the Poincaré flow only by terms of order $\mathcal{O}(\varepsilon^2)$. Due to the non-vanishing Θ component at $H' = 0$, the intersection yields a heteroclinic orbit of two saddles with Θ values differing by $\mathcal{O}(\varepsilon)$. In terms of the original variables $y_1 = (\varepsilon/a)^2 v$, $y_2 = (\varepsilon/a)^2 \lambda$ we therefore obtain a cusp of saddles in the equilibrium y-plane, with the two half-arcs connected almost horizontally by a heteroclinic orbit. Indeed the horizontal width of the heteroclinic sector is of order ε^3 at distance ε^2 from the origin, see Fig. 10.1.

We now have established the validity of the (10.24) not only on the Poincaré section Σ given by $|\tilde{H}| < \frac{2}{3}\sqrt{2}$, $\Theta > 0$, but up to the boundary $\tilde{H} = \frac{2}{3}\sqrt{2}$ of the saddle equilibria $v < 0$ and their strong stable and unstable manifolds $W^{ss}(v)$, $W^{uu}(v)$. We take another look at the phase portrait of (10.52), where we have replaced Θ by $\tau = \log(\Theta((b+1)/\lambda)^2)$, see Fig. 10.4. Note, however, that a time ε discretization step of (10.24) as realized by the Poincaré return map Π^ε corresponds to a τ-step of size

$$\Theta^{-1}\dot{\Theta}\varepsilon = \tfrac{2}{5}\Theta^{1/4}(b+1)(2J_1 + 3\tilde{H}J_0)\,\varepsilon. \tag{10.62}$$

Near the saddle boundary this expression simplifies to a τ-step of

$$\Theta^{-1}\dot{\Theta}\varepsilon = \tfrac{48}{5}(2\Theta)^{1/4}(b+1)\,\varepsilon, \tag{10.63}$$

by use of (10.39).

In the simplest cases, $(B), (C)$ with $\lambda < 0$, all orbits of the Poincaré flow are pointing strictly downwards. Therefore, the families

$$W^{cu} = \bigcup_{v<0} W^{uu}(v) \tag{10.64}$$

of strong unstable manifolds of the saddles intersects Σ in an infinite sequence of "horizontal" lines, accumulating to the stable foci along $\tilde{H} = -\frac{2}{3}\sqrt{2}$. All points of Σ in between these lines leave the neighborhood U of the Bogdanov-Takens point $x = y = 0$, in backwards time, while converging to the stable foci in forward time. We can view W^{cu} as a two-dimensional scroll, forward spiraling into the equilibria $x = 0, v > 0$, see Fig. 10.5.

The two most interesting cases are (B) and (C) with $\lambda > 0$ which exhibit Hopf bifurcation, without parameters of course, in coexistence with saddle heteroclinic orbits $\dot{\tilde{H}} = 0$ at $\tilde{H} = \frac{2}{3}\sqrt{2}$ as discussed before. The remaining two cases, (A) with $\lambda > 0$ and $\lambda < 0$, are similar to cases (B) and (C) with $\lambda > 0$, respectively: the former preserves the hyperbolic Hopf point but the heteroclinic orbit has disappeared through $\tau = +\infty$. Conversely, the latter preserves that heteroclinic orbit but has pushed the elliptic Hopf point out through $\tau = +\infty$, see Fig. 10.4

Case (C) with $\lambda > 0$, involves both a saddle heteroclinic orbit and an elliptic Hopf point; see Fig. 10.6. Locally near the elliptic Hopf point, Chap. 5 applies; this includes exponentially small splittings of heteroclinic separatrices, Fig. 5.1b. We first follow the unstable manifold $W^u(S)$ of the saddle segment S, between the two

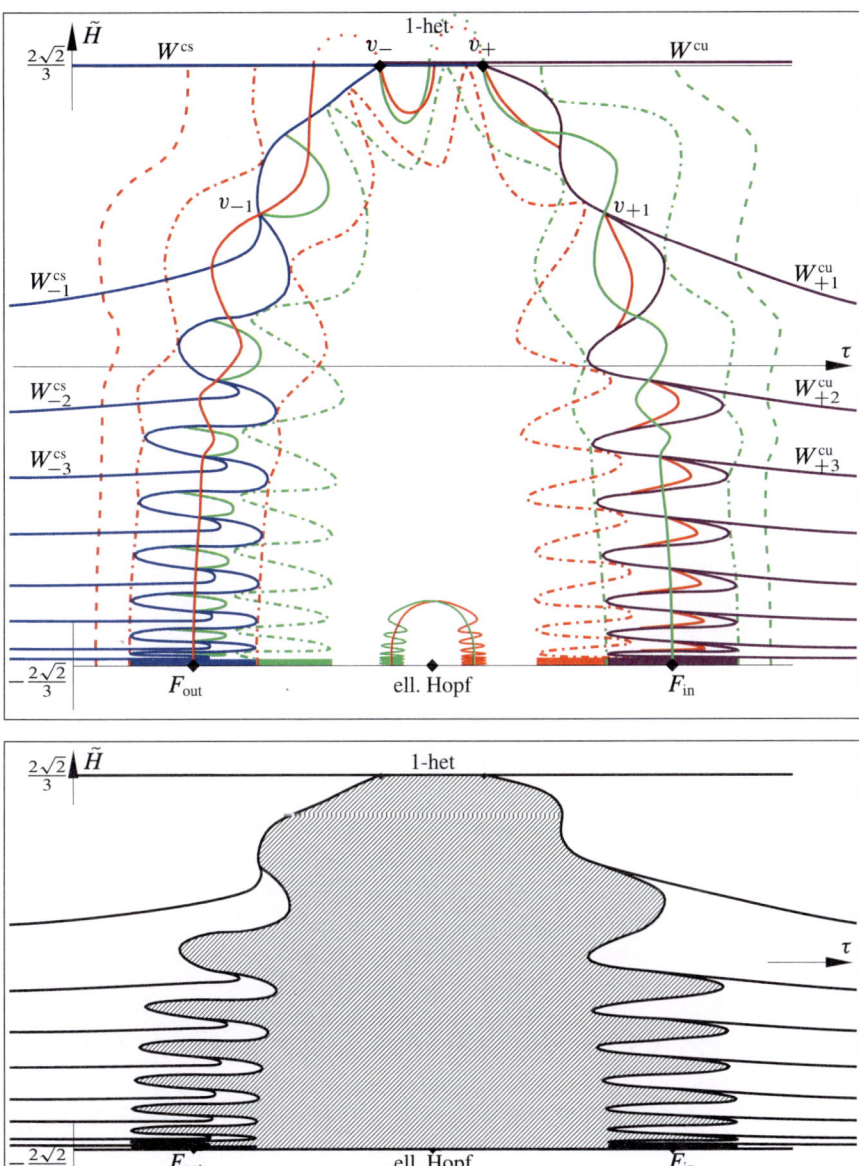

Fig. 10.6 Bogdanov-Takens point, Poincaré map, case (C), $\lambda > 0$. *Upper part*: Intersections of stable/unstable manifolds with the Poincaré section. Compare with Fig. 10.4. Color coding of manifolds: *magenta* $= W^{\mathrm{cu}}$(saddles), *blue* $= W^{\mathrm{cs}}$(saddles), *red* $= W^{\mathrm{uu}}$(focus), and *green* $= W^{\mathrm{ss}}$(focus). *Lower part*: Set of bounded orbits in the Poincaré section

saddle end points $v_\pm < 0$ of the saddle-saddle heteroclinic, forward under Π^ε. The (magenta) forward continuation is a piecewise smooth curve with tangent jumps of order ε at the (forward) points v_{+n}, $n = 0, 1, 2, \ldots$ of $W^u(v_+) \cap \Sigma$. Denote the limit by

$$F_{\text{in}} := \lim_{n \to \infty} v_{+n}$$

Typically the (green) stable manifold $W^{ss}(F_{\text{in}})$ will intersect $W^u(S)$ transversely, even at the points v_{+n}. The saddle segment S will therefore connect in forward time to an interval of stable equilibria around F_{in} to the right of the Hopf point. Note however, that transversality has not been proved but can typically be expected to hold for a first-order time-ε discretization of a flow. Similar statements hold true for the (blue) backwards continuation of of S in Σ and its intersection with the (red) unstable manifold of $W^{uu}(F_{\text{out}})$.

Figure 10.6 also illustrates the behavior of some strong stable and strong unstable manifolds of other foci $v > 0$ on the bottom line $\tilde{H} = -\frac{2}{3}\sqrt{2}$. Note how these manifolds transversely connect to equilibrium intervals on the other side of the elliptic Hopf point, or disappear partially, or disappear completely as their source points $v > 0$ move away from the Hopf point through $F_{\text{in/out}}$. Also, transverse splitting effects should not be expected to be exponentially small any more, during this transition, but to be of order ε. Outside the "Hopf bubble", we encounter the returns of the center-unstable and center-stable manifolds $W^{cu} \cap \Sigma$, $W^{cs} \cap \Sigma$ of the saddles to the right and left of S, respectively. These extend from the intersection points $v_{\pm n}$ and form smooth continuations of the piece of $W^u(S)$ immediately above $v_{\pm n}$.

Again we note that all non-stationary orbits that remain in the neighborhood are heteroclinic, as indicated in Fig. 10.1, with only one saddle-saddle heteroclinic, or also become unbounded through the split homoclinic family. These facts persist under higher-order perturbations.

The final case (B) with $\lambda > 0$ involves both a saddle heteroclinic orbit and a hyperbolic Hopf point; see Fig. 10.7. Locally, near the hyperbolic Hopf point, again Chap. 5 applies, see also Fig. 5.1a. This local analysis shows that the Hopf point itself possesses a stable (and an unstable) manifold W^{ss}_{Hopf}, (W^{uu}_{Hopf}) indicated in green (red) and hitting the saddle points S_{out}, S_{in} to the right (left) of the Hopf point. Note that most trajectories in W^{ss}_{Hopf}, (W^{uu}_{Hopf}) miss the line of saddles and leave the neighborhood in backward (forward) time. Strong stable (unstable) manifolds of further foci $v > 0$ at $\tilde{H} = -\frac{2}{3}\sqrt{2}$ are indicated. Also note the heteroclinic saddle-saddle connection from v_- to v_+ which has been established before.

From our analysis of the Poincaré flow, Fig. 10.4 (B), we conclude that Π^ε maps the collection W^{cu} of strong unstable manifolds of saddles to the right of v_- as is indicated by W^{cu}_{+n} in Fig. 10.7. Note how these manifolds W^{cu}_{+n} converge to the union of W^u_{Hopf} with the bottom line of normally stable equilibria to the right of the Hopf point. A similar pattern W^{cs}_{-n} arises from the center stable manifold W^{cs} of saddles to the left of v_+ under backwards iteration of Π^ε.

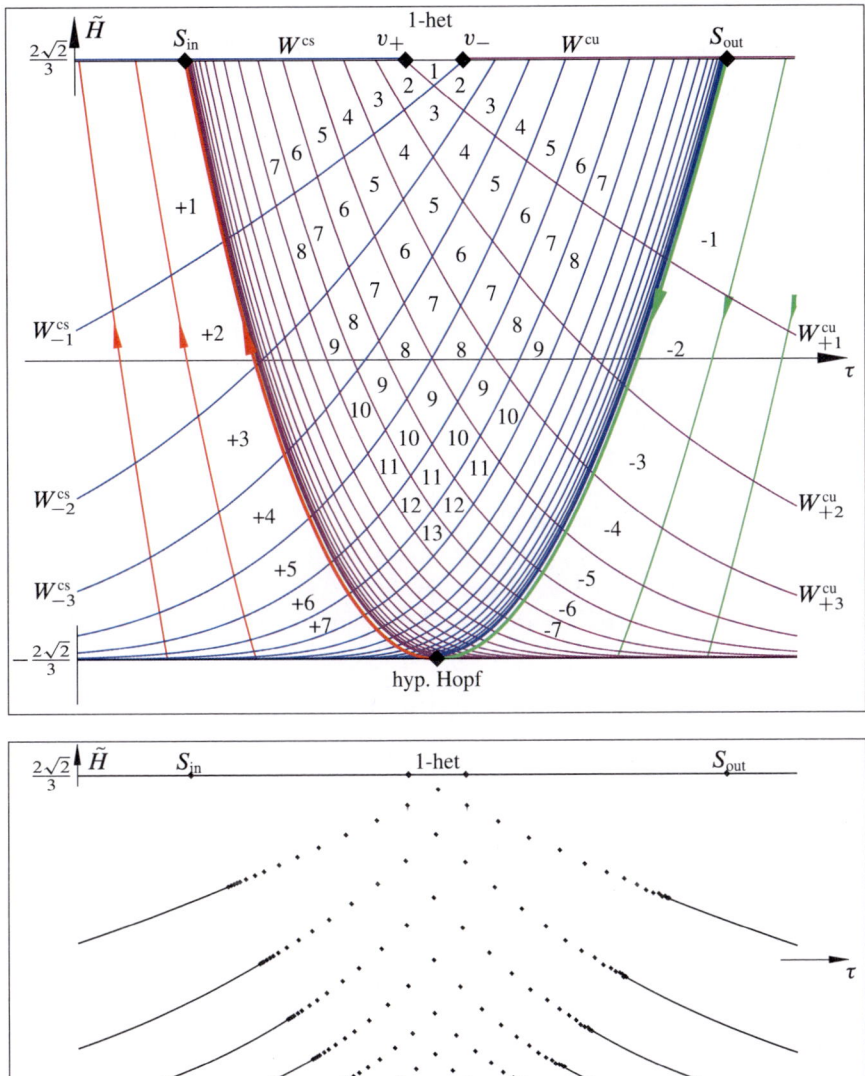

Fig. 10.7 Bogdanov-Takens point, Poincaré map, case (B), $\lambda > 0$. *Upper part*: Intersections of stable/unstable manifolds with the Poincaré section. Compare with Fig. 10.4. Color coding of manifolds: *magenta* $= W^{cu}$(saddles), *blue* $= W^{cs}$(saddles), *red* $= W^{uu}$(Hopf/focus), and *green* $= W^{ss}$(Hopf/focus). *Lower part*: Set of bounded orbits in the Poincaré section

By continuity of the curves W^{cu}_{+n}, $W^{\text{cs}}_{-n'}$, $n, n' = 0, 1, 2, \ldots$, each curve W^{cu}_{+n} must intersect each curve W^{cs}_{-n} at least once, in the sector of Σ between $W^{\text{uu}}_{\text{Hopf}}$ and $W^{\text{ss}}_{\text{Hopf}}$, say at a point $y^{n+n'}_n$. Then the $(n + n' + 1)$ points $y^{n+n'}_k$, $k = 0, \ldots, n + n'$, lie on an $(n + n')$-*heteroclinic* orbit from the saddle $y^{n+n'}_0$ to the saddle $y^{n+n'}_{n+n'}$. Indeed $\Pi^\varepsilon(y^{n+n'}_k) = y^{n+n'}_{k+1}$. Note $v_- = v^1_0$, $v_+ = y^1_1$, in this notation. We call these points $(n + n')$-heteroclinic because they revolve around the equilibrium line $x_1 = x_2 = 0$ for $(n + n')$ times before returning to the saddle line. This "heteroclinic swarm" is indicated as "n-het" in Fig. 10.1b. Note that neither (expected) uniqueness nor transversality of these infinitely many saddle-saddle heteroclinic orbits was addressed here. We have only established their existence.

In conclusion, except for equilibria and the above $(n + n')$-heteroclinic orbits $y^{n+n'}_k$, all points in the sector of Σ between $W^{\text{uu}}_{\text{Hopf}}$ and $W^{\text{ss}}_{\text{Hopf}}$ leave the region U in forward and backward time, due to homoclinic splitting, see the bottom part of Fig. 10.7. In the top part of Fig. 10.7 we have indicated lifetimes in numbers of revolution. Outside the sector we see a similar behavior to the region outside the Hopf bubble in the previous case (B).

Again we note that all non-stationary orbits that remain in the neighborhood are heteroclinic, as indicated in Fig. 10.1, including saddle-saddle $(n + n')$-heteroclinic orbits $y^{n+n'}_k$, $k = 0, \ldots, n + n'$, for any $n + n' = 1, 2, 3, \ldots$. These facts persist under higher-order perturbations.

Chapter 11
Zero-Hopf Bifurcation

In this chapter we study a bifurcation characterized by a zero eigenvalue and a pair of nonzero purely imaginary eigenvalues of the linearization transverse to a plane of equilibria. It turns out that instead we can study a one-parameter family of lines in a system depending on one parameter. Indeed, the rescaled normal form (11.6) is the same in both cases.

Consider a system

$$\begin{pmatrix} \dot{x} \\ \dot{y} \end{pmatrix} = F(x, y) = \begin{pmatrix} f(x, y) \\ g(x, y) \end{pmatrix}, \qquad x \in \mathbb{R}^3, \quad y \in \mathbb{R}^2, \tag{11.1}$$

$x = (x_1, x_2, x_3)$, $y = (y_1, y_2)$, $f = (f_1, f_2, f_3)$, $g = (g_1, g_2)$, with the following properties:

(i) The y-plane consists of equilibria, $F(0, y) \equiv 0$.
(ii) At the origin, the linearization takes the form

$$DF(0,0) = \begin{pmatrix} 0 & -1 & 0 & 0 & 0 \\ 1 & 0 & 0 & 0 & 0 \\ 0 & 0 & 0 & 0 & 0 \\ 0 & 0 & 1 & 0 & 0 \\ 0 & 0 & 0 & 0 & 0 \end{pmatrix}.$$

(iii) The critical eigenvalues cross the imaginary axis transversely:

$$\partial_{y_1} \operatorname{div}_{x_{1,2}} f_{1,2}(0,0) \neq 0,$$
$$\partial_{y_1} \partial_{x_3} f_3(0,0) \neq 0,$$
$$\nabla_y \operatorname{div}_{x_{1,2}} f_{1,2}(0,0) \nparallel \nabla_y \partial_{x_3} f_3(0,0).$$

© Springer International Publishing Switzerland 2015
S. Liebscher, *Bifurcation without Parameters*, Lecture Notes in Mathematics 2117,
DOI 10.1007/978-3-319-10777-6_11

(iv) We impose an additional non-degeneracy condition

$$\Delta_{x_{1,2}} f_3(0,0) \neq 0.$$

Note that (ii) is the generic form of a (suitably rescaled) linearization with one zero eigenvalue and one purely imaginary pair of eigenvalues along a plane of equilibria. Indeed, we take x_1, x_2 as the generalized real eigenvectors to the purely imaginary pair and x_3 as the eigenvector to the zero eigenvalue. We rescale time to normalize the imaginary eigenvalue and obtain the upper part $D_x f(0,0)$ of (ii). The kernel of $DF(0,0)$ has dimension at least 2, due to the plane of equilibria. Thus, generically, the kernel has dimension 2. Then the image of $DF(0,0)$ has dimension 3 and its intersection with the y-plane is one-dimensional and invariant under $DF(0,0)$. We take y_2 orthogonal to image$DF(0,0) \cap \{x = 0\}$. Hence

$$DF(0) = \begin{pmatrix} 0 & -1 & 0 & 0 & 0 \\ 1 & 0 & 0 & 0 & 0 \\ 0 & 0 & 0 & 0 & 0 \\ c_1 & c_2 & 1 & 0 & 0 \\ 0 & 0 & 0 & 0 & 0 \end{pmatrix}. \tag{11.2}$$

The shear $\tilde{y}_1 = y_1 + c_2 x_1 - c_1 x_2$ yields (ii).

Due to (iii) and the implicit-function theorem, there exist a curve of Poincaré-Andronov-Hopf points and a curve of transcritical points in the y plane. Both curves intersect transversely at the origin. A suitable shear transformation

$$\tilde{x}_3 = c_1 x_3, \qquad \tilde{y}_1 = c_1 y_1 + c_2 y_2, \qquad \tilde{y}_2 = c_3 y_2.$$

preserves the linearization and normalizes

$$\nabla_y \mathrm{div}_{x_{1,2}} f_{1,2}(0,0) = \varrho \begin{pmatrix} 1 \\ -1 \end{pmatrix}, \qquad \nabla_y \partial_{x_3} f_3(0,0) = \begin{pmatrix} 1 \\ 0 \end{pmatrix}, \tag{11.3}$$

with real coefficient $\varrho \neq 0$. Then the curve of transcritical points is tangential to the y_2-axis, and the curve of Hopf points is tangential to the diagonal $y_1 = y_2$. Both are still transverse to the linear drift direction y_2.

The normal-form procedure, see Sects. 2.2, 2.3, and [72], yields a normal form with additional rotational equivariance:

$$\begin{aligned} \dot{r} &= r h_r(r^2, x_3, y) + \text{h.o.t.}, \\ \dot{\varphi} &= h_\varphi(r^2, x_3, y) + \text{h.o.t.}, \\ \dot{x}_3 &= h_3(r^2, x_3, y) + \text{h.o.t.}, \\ \dot{y}_1 &= x_3 + h_1(r^2, x_3, y) + \text{h.o.t.}, \\ \dot{y}_2 &= h_2(r^2, x_3, y) + \text{h.o.t.}. \end{aligned} \tag{11.4}$$

Polynomials h, in normal form, do not depend on the angle φ. Terms of higher order, beyond normal form, depend on all variables and generically break the normal-form symmetry.

The plane of equilibria and the linearization at the origin are preserved by the normal form procedure, thus $h_r(0,0,0) = 0$, $h_\varphi(0,0,0) = 1$, $h_k(0,0,y) \equiv 0$, due to conditions (i) and (ii). The multiplier $1/\dot{\varphi}$ is close to 1, preserves trajectories, and normalizes the rotation speed. Thus we can put $\dot{\varphi} = 1$ in (11.4). Condition (11.3) implies $\nabla_y h_r(0,0,0) = \varrho \begin{pmatrix} 1 \\ -1 \end{pmatrix}$ and $\nabla_y \partial_{x_3} f_3(0,0) = \begin{pmatrix} 1 \\ 0 \end{pmatrix}$.

We drop the ϕ component and rescale the system by

$$
\begin{aligned}
r &= \sigma^3 \tilde{r}, \\
x_3 &= \sigma^4 \tilde{u}, \\
y_1 &= \sigma^2 \tilde{v}, \\
y_2 &= \sigma^2 \tilde{\lambda}, \\
t &= \sigma^{-2} \tilde{t}.
\end{aligned}
\tag{11.5}
$$

For $0 < \sigma \ll 1$, to leading order in σ, we obtain the rescaled normal form

$$
\begin{aligned}
\dot{r} &= \varrho(v - \lambda)r + r\mathcal{O}(\sigma), \\
\dot{u} &= uv + ar^2 + \mathcal{O}(\sigma), \\
\dot{v} &= u + \mathcal{O}(\sigma), \\
\dot{\lambda} &= \mathcal{O}(\sigma).
\end{aligned}
\tag{11.6}
$$

Note the renaming of variables to simplify notation and to emphasize the role of $y_2 = \sigma^2\lambda$ as a parameter of the truncated rescaled normal form. As we remarked in the beginning of this chapter, a Zero-Hopf point on a plane of equilibria without parameters and a Zero-Hopf on a line of equilibria with additional parameter both result in the same rescaled normal form (11.6).

Note further that (11.6) is also the normal form for a crossing of a transcritical point and another transcritical point with $\mathbb{Z}_2$ equivariance.

System (11.6) displays a line of transcritical points for $r = u = v = 0$ and a line of Poincaré-Andronov-Hopf points for $r = u = 0, v = \lambda$. The coefficient $\varrho \neq 0$ can be interpreted as the ratio of the crossing speeds of the Hopf and the transcritical eigenvalues through the imaginary axis, at $v = \lambda = 0$ as v is varied.

In (11.6) we can normalize $\lambda = +1$, by the scaling $v = \lambda\tilde{v}$, $u = \lambda^2\tilde{u}$, $t = \lambda^{-1}\tilde{t}$, Condition (iv) ensures $a \neq 0$, and we can normalize $a = \pm 1$, by scaling of r. We finally arrive at the truncated normal form

$$
\begin{aligned}
\dot{r} &= \varrho(v - 1)r, \\
\dot{u} &= uv + ar^2, \\
\dot{v} &= u,
\end{aligned}
\tag{11.7}
$$

with $\varrho \neq 0$ and $a = \pm 1$. In this normal-form flow, we find the v-axis of equilibria, a transcritical point at the origin with critical eigenvector u, and a transcritical point with reflection symmetry—alias a Poincaré-Andronov-Hopf point—at $r = 0$, $u = 0$, $v = 1$. The absolute value $|\varrho|$ is the ratio of the speeds of the transverse eigenvalue crossings of the Hopf and the zero eigenvalue.

Let us verify the Hopf point and determine its type. The linearization at $r = 0$, $u = 0$, $v = 1$ is

$$\begin{pmatrix} 0 & 0 & 0 \\ 0 & 1 & 0 \\ 0 & 1 & 0 \end{pmatrix},$$

with kernel vectors $(1, 0, 0)$ and $(0, 0, 1)$ and unstable eigenvector $(0, 1, 1)$. The projection

$$\Pi_c = \begin{pmatrix} 1 & 0 & 0 \\ 0 & 0 & 0 \\ 0 & 1 & -1 \end{pmatrix}$$

of (11.7) onto the center eigenspace at $v = 1$ yields the reduced vector field on the center manifolds to second order

$$\begin{aligned} \dot{\delta}_r &= \dot{r} &= \varrho \delta_v \delta_r, \\ \dot{\delta}_v &= \dot{u} - \dot{v} = u(\delta_v + 1) + ar^2 - u &= a\delta_r^2. \end{aligned} \tag{11.8}$$

Here $(\delta_r, \delta_v) = (r, v - 1)$ are the local coordinates on the center eigenspace. Note that a nonlinear expansion of the center manifold is not needed to determine the reduced vector field to second order.

We compare (11.8) with Chap. 5 and Sect. 4.2 to find a hyperbolic Hopf point, for $\varrho a > 0$, and an elliptic Hopf point, for $\varrho a < 0$. The 4 main cases are shown in Fig. 11.1.

Note that $v(t)$ is almost a Lyapunov function for the normal-form flow (11.7). Indeed, if $u > 0$ then v strictly increases. If $u < 0$, then v strictly decreases. However u cannot cross zero more than once, and the crossing direction is determined by the sign of a: $\dot{u}|_{u=0} = ar^2$ has fixed nonzero sign outside the line $r = u = 0$ of equilibria. In particular, we conclude:

Remark 11.1 Given (11.1) with conditions (i–iv). Then the set of small bounded trajectories near the bifurcation point at the origin consists of the given plane of equilibria and of heteroclinic orbits between them.

Let us study the case $a = -1$, $0 < \varrho$ of (11.7) in more detail. The line of equilibria is normally stable for $v < 0$, and normally stable in reversed time for $v > 1$. The Hopf point $v = 1$ is elliptic.

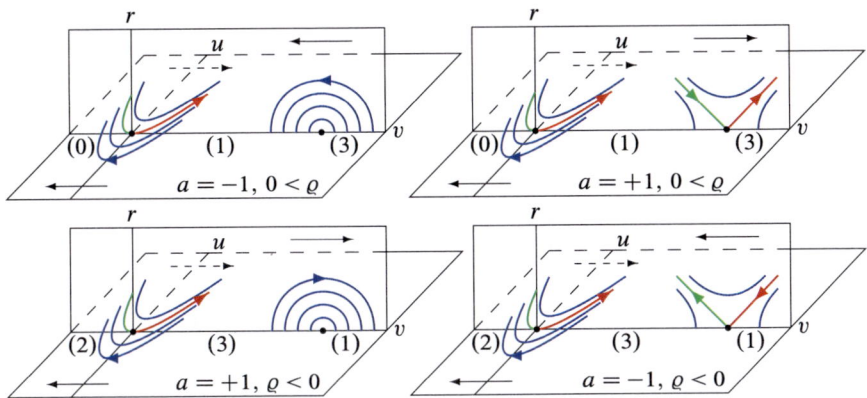

Fig. 11.1 Cases of Zero-Hopf bifurcation. The pictures show the 4 main cases of Zero-Hopf bifurcation (11.7). Note the relative position of transcritical and Hopf point, the unstable dimension (*u*) of the equilibria, and the drift directions in the coordinate planes. Compare with Figs. 4.1b and 4.3, 5.1

Lemma 11.2 *Let* $a = -1$ *and* $0 < \varrho$. *Orbits* $(r, u, v)(t)$ *starting for* $t = 0$ *in the half plane* $\{u = 0, r > 0\}$ *converge to the line of equilibria for* $t \to \infty$.

Proof Orbits $(r, u, v)(t)$ starting for $t = 0$ in the half plane $\{u = 0, r > 0\}$ cross the plane transversely, $\dot{u}(0) = ar^2 < 0$, and stay in $\{u < 0\}$ for all $t > 0$. Therefore $v(t)$ is a strict Lyapunov function for $t > 0$.

If $v(t)$ is bounded, then it converges: $\lim_{t \to \infty} v(t) = v_\infty$. This implies the vanishing limits $\lim_{t \to \infty} u(t) = 0$ and $\lim_{t \to \infty} r(t) = 0$ due to (11.7). Thus, the limit is an equilibrium as claimed.

Therefore, assume that $\lim_{t \to \infty} v(t) = -\infty$. Then r decays to zero, in fact with decay rate $\varrho(v - 1) \to -\infty$. As soon as r is small enough, u decay to zero. The equilibria $v < 0$ are normally stable, thus $r, u \to 0$ implies convergence to a single equilibrium. This is a contradiction to the assumption $v(t) \to -\infty$. □

This convergence result holds true for the full system (11.1) in a small enough neighborhood of the bifurcation point. Indeed, transversal crossing persists under perturbation. Furthermore, the normal-form flow only needs finite time to enter the the domain of attraction of the equilibria. Thus, the perturbed flow will also enter the domain of attraction.

Remark 11.3 Let $a = -1$ and $0 < \varrho$. The strong stable local manifolds $W^{ss}_{loc}(v)$ to equilibria $v \approx 0$ near the transcritical point forms a manifold tangential to the (r, v)-plane. It is the unique sets of orbits which converge to the equilibria and are tangent to the (r, v)-plane. Therefore $W^{ss}(v) \subset \{u > 0\}$. In particular, along trajectories on $W^{ss}(v)$, the component v strictly increases.

Proof Orbits $z(t) = (r, u, v)(t)$ starting for $t = 0$ in the half plane $\{u = 0, r > 0\}$ cross the plane transversely, $\dot{u}(0) = ar^2 < 0$. Then they stay in $\{u < 0\}$ for all $t > 0$ and converge to the equilibrium line.

Assume that a piece of W^{ss} would be contained in $\{u < 0\}$. Then the entire forward orbit of this piece must be contained in $\{u < 0\}$. But the forward orbit is also tangential to the (r, v)-plane. Therefore, orbits starting between W^{ss} and the (r, v)-plane converge to the line of equilibria and are also tangent to the (r, v)-plane. This contradicts the uniqueness of W^{ss}. □

Theorem 11.4 *Let $a = -1$ and $1/2 < \varrho$. Consider an arbitrary initial value $z(0) = (r_0, u_0, v_0)$ with $r_0 > 0$ to the normal-form system (11.7). Then the trajectory converges for $t \to \infty$ to an equilibrium $(0, 0, v_\infty)$. For $0 < \varrho < 1/2$ orbits may escape to infinity.*

Proof As soon as u becomes non-positive, lemma 11.2 yields the claim. We assume $u(0) > 0$. The r component stays positive for all time. We obtain

$$
\begin{aligned}
\frac{d}{dt}\left(\frac{u}{r^2}\right) &= \frac{1}{r^4}\left(\dot{u}r^2 - 2ur\dot{r}\right) \\
&= \frac{1}{r^4}\left(uvr^2 - r^4 - 2\varrho(v-1)ur^2\right) \\
&= \frac{u}{r^2}\left(v - 2\varrho(v-1)\right) - 1.
\end{aligned}
\tag{11.9}
$$

Assume that u stays positive. Then v strictly increases. It it converges to a limit, then the solution converges to an equilibrium, due to the same arguments as in the proof of lemma 11.2. Assume, on the other hand, that v is unbounded. Then $v - 2\varrho(v-1)$ becomes and stays negative, provided $1/2 < \varrho$. (If $0 < \varrho < 1/2$ and v is large enough then $v - 2\varrho(v-1)$ stays positive and the trajectory escapes.) Therefore, u/r^2 eventually becomes negative. Hence, u cannot stay positive. □

For $\rho > 1/2$, that is if the transverse crossing of the Hopf eigenvalue pair is fast enough compared to the crossing speed of the transcritical simple zero eigenvalue, then there is no escape in forward time, except on the singular boundary $r = 0$. Although the manifold of equilibria becomes normally unstable at the bifurcation points, all trajectories which are repelled from the unstable region of the manifold are recovered by the stable side. Geometrically, the elliptic bubble emerging from the Hopf point extends to a cusp shaped domain touching the saddles $v < 0$ from the negative u direction.

Remark 11.5 Theorem 11.4 holds true for the case $a = +1$ and the reversed flow, by an analogous calculation. There is no escape for $1/2 < \varrho$ in backward direction. Escape is possible for $0 < \varrho < 1/2$.

For the other two cases, $\varrho < 0$, sources and sinks (with unstable dimension 3 and 0) do not appear simultaneously, thus heteroclinic orbits do not fill open regions.

Taking terms of higher order into account, the open regions of heteroclinic orbits persist. The escape on the singular boundary, for the normal form, could induce an open region of escaping trajectories: higher-order terms could drive orbits towards the boundary. Further research is necessary to get more refined results.

Chapter 12
Double-Hopf Bifurcation

The final bifurcation of codimension 2 is characterized by the intersection of 2 curves of Poincaré-Andronov-Hopf points on a two-dimensional surface of equilibria. As we shall see, the drift direction at the Hopf lines play an important role. In the case of a parameter-dependent fixed line of equilibria, drifts at both Hopf-lines can be opposite and spiraling orbits appear, see Sect. 12.1. In the generic case with a plane of equilibria without parameters, both drifts are transverse and generate a Lyapunov function. Only heteroclinic orbits arise. See Sect. 12.2.

12.1 Family of Lines of Equilibria

Consider a system

$$
\left.\begin{array}{l}
\begin{pmatrix} \dot{x} \\ \dot{y} \end{pmatrix} = F(x, y, \lambda) = \begin{pmatrix} f(x, y, \lambda) \\ g(x, y, \lambda) \end{pmatrix} \\[6pt]
\dot{\lambda} \;= 0,
\end{array}\right\} \quad x \in \mathbb{R}^4, \quad y, \lambda \in \mathbb{R}, \qquad (12.1)
$$

$x = (x_1, x_2) = (x_{11}, x_{12}, x_{21}, x_{22})$, $f = (f_1, f_2)$, with the following properties:

(i) For all parameter values, there exists a line of equilibria, $F(0, y, \lambda) \equiv 0$, forming a plane of equilibria in the extended phase space.
(ii) The linearization at the origin possesses two pairs of purely imaginary eigenvalues with irrational quotient: w.l.o.g.

© Springer International Publishing Switzerland 2015
S. Liebscher, *Bifurcation without Parameters*, Lecture Notes in Mathematics 2117,
DOI 10.1007/978-3-319-10777-6_12

$$DF(0) = \begin{pmatrix} 0 & -1 & 0 & 0 & 0 & 0 \\ 1 & 0 & 0 & 0 & 0 & 0 \\ 0 & 0 & 0 & -\omega & 0 & 0 \\ 0 & 0 & \omega & 0 & 0 & 0 \\ 0 & 0 & 0 & 0 & 0 & 0 \\ 0 & 0 & 0 & 0 & 0 & \end{pmatrix}, \qquad \omega \in \mathbb{R} \setminus \mathbb{Q}.$$

(iii) The critical eigenvalues cross the imaginary axis transversely:

$$\partial_y \operatorname{div}_{x_1} f_1(0) \neq 0,$$
$$\partial_y \operatorname{div}_{x_2} f_2(0) \neq 0,$$
$$\nabla_{y,\lambda} \operatorname{div}_{x_1} f_1(0) \nparallel \nabla_{y,\lambda} \operatorname{div}_{x_2} f_2(0).$$

The implicit-function theorem then yields curves of Hopf-points orthogonal to $\nabla_{y,\lambda} \operatorname{div}_{x_1} f_1(0)$ and $\nabla_{y,\lambda} \operatorname{div}_{x_2} f_2(0)$.

(iv) The drift along the line of equilibria is non-degenerate:

$$0 \neq \Delta_{x_1} g(0) \neq \Delta_{x_2} g(0) \neq 0.$$

In particular, the aforementioned Hopf-points are generic outside the origin, and therefore of the form discussed in Chap. 5.

Given (iii), we can normalize

$$\nabla_{y,\lambda} \operatorname{div}_{x_1} f_1(0) = \begin{pmatrix} 1 \\ 0 \end{pmatrix}, \qquad \nabla_{y,\lambda} \operatorname{div}_{x_2} f_2(0) = a \begin{pmatrix} 1 \\ 1 \end{pmatrix}, \qquad (12.2)$$

with $a \neq 0$.

The normal-form procedure, see Sects. 2.2, 2.3, and [72], yields a normal form with additional equivariance with respect to rotations by $\{(\alpha, \omega\alpha); \alpha \in \mathbb{R}\}$ in (x_1, x_2). We write $x_k = r_k \exp(i\phi_k)$, $k = 1, 2$ in polar coordinates. Due to the irrationality of ω this group of rotations is dense on the torus $S^1 \times S^1$ and the normal form is independent of both angles ϕ_1, ϕ_2:

$$\begin{aligned}
\dot{r}_1 &= r_1 h_{r_1}(r_1^2, r_2^2, y, \lambda) + \text{h.o.t.}, \\
\dot{\phi}_1 &= 1 + h_{\phi_1}(r_1^2, r_2^2, y, \lambda) + \text{h.o.t.}, \\
\dot{r}_2 &= r_2 h_{r_1}(r_1^2, r_2^2, y, \lambda) + \text{h.o.t.}, \\
\dot{\phi}_2 &= \omega + h_{\phi_2}(r_1^2, r_2^2, y, \lambda) + \text{h.o.t.}, \\
\dot{y} &= h_{y_1}(r_1^2, r_2^2, y, \lambda) + \text{h.o.t.},
\end{aligned} \qquad (12.3)$$

Polynomials h, in normal form, do not depend on the angle φ. Terms of higher order, beyond normal form, depend on all variables and generically break the normal-form symmetry. Conditions (i–iv) and (12.2) imply

$$
\begin{aligned}
\dot{r}_1 &= r_1 y + \mathcal{O}(\|z\|^3), \\
\dot{r}_2 &= a r_2 (y + \lambda) + \mathcal{O}(\|z\|^3), \\
\dot{y} &= b_1 r_1^2 + b_2 r_2^2 + \mathcal{O}(\|z\|^3),
\end{aligned}
\tag{12.4}
$$

with $z = (r_1, r_2, y, \lambda)$. Coefficients a, b_1, b_2 are nonzero. We can normalize $\lambda = -1$ by scaling of λ, y and time. Then, by scaling of b_1, b_2, we can normalize $r_1 = \pm 1$ and $r_2 = \pm 1$. The final truncated normal form reads

$$
\begin{aligned}
\dot{r}_1 &= r_1 y, \\
\dot{r}_2 &= a r_2 (y - 1), \\
\dot{y} &= b_1 r_1^2 + b_2 r_2^2,
\end{aligned}
\tag{12.5}
$$

width $a \neq 0$, $b_1 = \pm 1$, $b_2 = \pm 1$.

The drifts b_1, b_2 can be of the same or of opposite direction, and the Hopf points can be both elliptic, both hyperbolic, or one of each type. These are six main cases.

Consider, for example the case $b_1 = -1$, $b_2 = 1$, $a < 0$ of 2 elliptic Hopf points with opposite drift. Then the distances

$$
d_1 = y^2 + r_1^2 - \tfrac{1}{a} r_2^2, \qquad d_2 = (y - 1)^2 + r_1^2 - \tfrac{1}{a} r_2^2
$$

from the two Hopf points monotonically increase,

$$
\dot{d}_1 = 2 r_2^2, \qquad \dot{d}_2 = 2 r_1^2,
$$

outside the singular boundary $\{r_1 r_2 = 0\}$. Furthermore,

$$
\tfrac{d}{dt}(r_1 r_2) = [(1 + a) y - 1] r_1 r_2.
$$

If $a \approx -1$, i.e. if the both pairs of Hopf eigenvalues cross the imaginary axis as approximately the same speed, then $r_1 r_2$ decreases near the bifurcation point. Trajectories approach the singular boundary $\{r_1 r_2 = 0\}$ while alternately following the heteroclinic connections of the two elliptic Hopf bubbles.

In Bianchi models, Chap. 13, the flow near Taub exhibits a singular version of this flow. There, additional symmetries yield $a = -1$ and $\lambda = 0$, i.e. the unfolding parameter is missing.

A detailed analysis of this normal form has not yet been carried out.

12.2   Plane of Equilibria

Consider a system

$$
\begin{pmatrix} \dot{x} \\ \dot{y} \end{pmatrix} = F(x, y) = \begin{pmatrix} f(x, y) \\ g(x, y) \end{pmatrix} \left. \right\} \quad x \in \mathbb{R}^4, \quad y \in \mathbb{R}^2, \qquad (12.6)
$$
$$
\dot{\lambda} = 0,
$$

$x = (x_1, x_2) = (x_{11}, x_{12}, x_{21}, x_{22})$, $f = (f_1, f_2)$, $y = (y_1, y_2)$, $g = (g_1, g_2)$, with the following properties:

(i) There exists a plane of equilibria, $F(0, y) \equiv 0$.
(ii) The linearization at the origin possesses two pairs of purely imaginary eigenvalues with irrational quotient: w.l.o.g.

$$
DF(0) = \begin{pmatrix}
0 & -1 & 0 & 0 & 0 & 0 \\
1 & 0 & 0 & 0 & 0 & 0 \\
0 & 0 & 0 & -\omega & 0 & 0 \\
0 & 0 & \omega & 0 & 0 & 0 \\
0 & 0 & 0 & 0 & 0 & 0 \\
0 & 0 & 0 & 0 & 0 &
\end{pmatrix}, \qquad \omega \in \mathbb{R} \setminus \mathbb{Q}. \qquad (12.7)
$$

(iii) The critical eigenvalues cross the imaginary axis transversely, and the drift along the plane of equilibria is non-degenerate: every two of the following vectors in $\mathbb{R}^2$ are transverse

$$
\nabla_y \mathrm{div}_{x_1} f_1(0), \qquad \nabla_y \mathrm{div}_{x_2} f_2(0), \qquad \Delta_{x_1} g(0), \qquad \Delta_{x_2} g(0),
$$

i.e. all vectors are nonzero and none are parallel. The implicit-function theorem then yields curves of Hopf-points which are orthogonal to $\nabla_{y,\lambda} \mathrm{div}_{x_1} f_1(0)$ and $\nabla_{y,\lambda} \mathrm{div}_{x_2} f_2(0)$.

Given (iii), we can take $\Delta_{x_1} g(0)$ and $\Delta_{x_2} g(0)$ as new coordinates y_1, y_2. Therefore, we assume w.l.o.g. that

$$
\Delta_{x_1} g(0) = \begin{pmatrix} 1 \\ 0 \end{pmatrix}, \qquad \Delta_{x_2} g(0) = \begin{pmatrix} 0 \\ 1 \end{pmatrix}. \qquad (12.8)
$$

The transversality of these two drift direction is also the main difference to the mixed case of Sect. 12.1 where both drifts were restricted to the y_1 direction and therefore parallel by definition.

The normal-form procedure, see Sects. 2.2, 2.3, and [72], yields a normal form with additional equivariance with respect to rotations by $\{(\alpha, \omega\alpha); \alpha \in \mathbb{R}\}$ in (x_1, x_2). We write $x_k = r_k \exp(i\phi_k)$, $k = 1, 2$ in polar coordinates. Due to the irrationality of ω this group of rotations is dense on the torus $S^1 \times S^1$ and the normal form is independent of both angles ϕ_1, ϕ_2:

$$
\begin{aligned}
\dot{r}_1 &= r_1 h_{r_1}(r_1^2, r_2^2, y) + \text{h.o.t.}, \\
\dot{\phi}_1 &= 1 + h_{\varphi_1}(r_1^2, r_2^2, y) + \text{h.o.t.}, \\
\dot{r}_2 &= r_2 h_{r_1}(r_1^2, r_2^2, y) + \text{h.o.t.}, \\
\dot{\phi}_2 &= \omega + h_{\varphi_2}(r_1^2, r_2^2, y) + \text{h.o.t.}, \\
\dot{y}_1 &= h_{y_1}(r_1^2, r_2^2, y) + \text{h.o.t.}, \\
\dot{y}_2 &= h_{y_2}(r_1^2, r_2^2, y) + \text{h.o.t.}
\end{aligned}
\tag{12.9}
$$

Polynomials h, in normal form, do not depend on the angle φ. Terms of higher order, beyond normal form, depend on all variables and generically break the normal-form symmetry. Conditions (i–iii) and (12.8) imply

$$
\begin{aligned}
\dot{r}_1 &= r_1(a_{11}y_1 + a_{12}y_2) + \mathcal{O}(\|z\|^3), \\
\dot{r}_2 &= r_2(a_{21}y_1 + a_{22}y_2) + \mathcal{O}(\|z\|^3), \\
\dot{y}_1 &= r_1^2 + \mathcal{O}(\|z\|^3), \\
\dot{y}_2 &= r_2^2 + \mathcal{O}(\|z\|^3),
\end{aligned}
\tag{12.10}
$$

with $z = (r_1, r_2, y_1, y_2)$. All a_{kl} and the determinant $a_{11}a_{22} - a_{12}a_{21}$ are nonzero.

In particular, we find the Lyapunov function $V(z) = -(y_1 + y_2)$ strictly decreasing along trajectories except at equilibria $r_1 = r_2 = 0$. Again, recurrent dynamics does not arise.

Remark 12.1 Given (12.6) with conditions (i–iii). Then the set of small bounded trajectories near the bifurcation point at the origin consists of the given plane of equilibria and of heteroclinic orbits between them.

A detailed analysis of this normal form has not yet been carried out.

Chapter 13
Application: Cosmological Models of Bianchi Type, the Tumbling Universe

Cosmological models are solutions of the Einstein equations

$$\mathrm{Ric}(g) - \tfrac{1}{2}\mathrm{Scal}(g)g \; = \; T \tag{13.1}$$

relating the geometry of spacetime—the curvature of a four-dimensional Lorentzian metric g—to the matter content. The Einstein equations are usually coupled to kinetic equations of the matter.

Solving the full PDE system (13.1) is beyond anybody's abilities up to now. Therefore, additional symmetries are assumed to discuss special solutions.

The simplest model—and core of the standard model used to describe the evolution of our universe—is the Friedmann model of spatially homogeneous and isotropic spacetimes. This assumption of a six-dimensional symmetry group allows a reduction of (13.1) to one scalar ODE that determines the expansion rate of the universe. This expansion rate can be compared with the measurements of the Hubble constant and with the consequences of large-scale thermodynamics of the matter part.

The next step towards the full PDE system are Bianchi models of spatially homogeneous but anisotropic spacetimes. In other words, the spacetime is assumed to be foliated into spatial hypersurfaces given by the orbits of a three-dimensional symmetry group. In the simplest case of Bianchi class A, system (13.1) with perfect-fluid matter model can then be reduced to a five-dimensional ODE system in expansion-reduced variables,

$$
\begin{aligned}
N_1' &= (q - 4\Sigma_+)N_1, \\
N_2' &= (q + 2\Sigma_+ + 2\sqrt{3}\Sigma_-)N_2, \\
N_3' &= (q + 2\Sigma_+ - 2\sqrt{3}\Sigma_-)N_3, \\
\Sigma_+' &= (q - 2)\Sigma_+ - 3S_+, \\
\Sigma_-' &= (q - 2)\Sigma_- - 3S_-,
\end{aligned}
\tag{13.2}
$$

© Springer International Publishing Switzerland 2015
S. Liebscher, *Bifurcation without Parameters*, Lecture Notes in Mathematics 2117,
DOI 10.1007/978-3-319-10777-6_13

with the abbreviations

$$
\begin{aligned}
S_+ &= \tfrac{1}{2}\left((N_2 - N_3)^2 - N_1\,(2N_1 - N_2 - N_3)\right), \\
S_- &= \tfrac{1}{2}\sqrt{3}\,(N_3 - N_2)\,(N_1 - N_2 - N_3), \\
q &= 2\left(\Sigma_+^2 + \Sigma_-^2\right) + \tfrac{1}{2}(3\gamma - 2)\Omega, \\
\Omega &= 1 - \Sigma_+^2 - \Sigma_-^2 - K, \\
K &= \tfrac{3}{4}\left(N_1^2 + N_2^2 + N_3^2 - 2\,(N_1 N_2 + N_2 N_3 + N_3 N_1)\right).
\end{aligned}
\tag{13.3}
$$

This system is due to Wainwright and Hsu [74]. The initial big-bang singularity is approached in the limit time to $-\infty$. Variables N_k describe the curvature of spatial hypersurfaces. Their signs determine the Lie-algebra type of the associated spatial symmetry imposed by the homogeneity assumption. Due to Bianchi's classifications of three-dimensional Lie algebras—the tangent spaces to the assumed symmetry group—these models are called Bianchi models, although they have been introduced by Gödel and Taub. The variables $\Sigma_\pm$ relate to the second fundamental form of the spatial hypersurfaces. The matter density Ω is positive, the boundary $\Omega = 0$ is invariant. The coefficient $\gamma < 2$ describes the perfect fluid, e.g. $\gamma = 4/3$ for radiation and $\gamma = 1$ for dust. See also [73] for further details on this dynamics approach to cosmology and [40] for a review on current knowledge of Bianchi models and open questions.

Prominent feature of system (13.2) is the Kasner circle $\mathscr{K}$ of equilibria,

$$
\mathscr{K} = \{\ \Sigma_+^2 + \Sigma_-^2 = 1,\ N_1 = N_2 = N_3 = 0\ \},
\tag{13.4}
$$

and the caps

$$
\mathscr{K}_k^\pm = \{\ \Sigma_+^2 + \Sigma_-^2 = 1 - N_k^2,\ \pm N_k > 0\ N_{k+1} = N_{k-1} = 0,\ k \bmod 3\ \},
\tag{13.5}
$$

filled with heteroclinic orbits connecting equilibria on $\mathscr{K}$. The projection of the heteroclinic orbits to the Σ-plane lie on straight lines through the corners of a circumscribed triangle, see Fig. 13.1.

The Kasner circle itself is normally hyperbolic except at the Taub points

$$
T_1 = (-1, 0),\quad T_2 = (1/2, \sqrt{3}/2),\quad T_3 = (1/2, -\sqrt{3}/2),
\tag{13.6}
$$

in coordinates (Σ_+, Σ_-). At the Taub points, two nontrivial eigenvalues of the linearization at Kasner equilibria cross zero in opposite direction.

Thus, bifurcations without parameters arise. However, the system is not in "generic" position. In particular, additional equivariances

$$
\begin{aligned}
&\text{reflection} && (N, \Sigma) \mapsto (-N, \Sigma), \\
&\text{cyclic permutation } (N_1, N_2, N_3, \Sigma) \mapsto (N_2, N_3, N_1, e^{2\pi i/3}\,\Sigma)
\end{aligned}
\tag{13.7}
$$

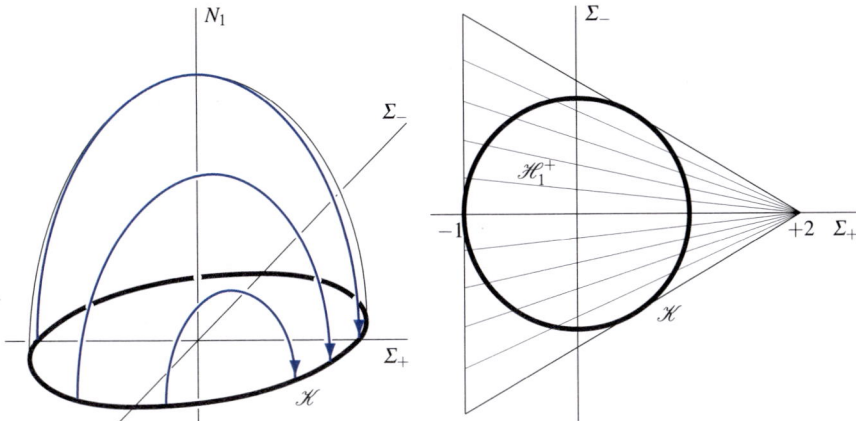

Fig. 13.1 Kasner circle $\mathscr{K}$ of equilibria and heteroclinic cap $\mathscr{H}_1^+$

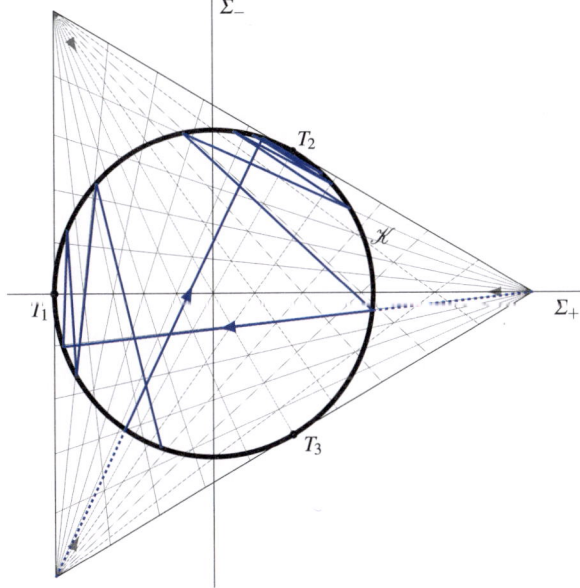

Fig. 13.2 Kasner map. Heteroclinic Bianchi solutions in reversed time direction towards the big-bang singularity

are inherited from the geometric origin of the model, see also Sect. 12.1. Furthermore, we are not interested in small bounded solutions near T_k. Rather than small solutions, passages near the Taub points and global re-entry into the neighborhood of another Taub point are of interest, see Fig. 13.2.

The task is to study trajectories following formal chains of heteroclinic orbits, given by the heteroclinic caps $\mathscr{H}_k$ and inducing (in backward direction) a non-uniformly expanding map of the Kasner circle $\mathscr{K}$ onto itself. This Kasner map is believed to govern the dynamics of the early universe, at least in the Bianchi model, close to the big-bang singularity $t \rightarrow -\infty$. The mixing properties of the formal shift dynamics of heteroclinic sequences are conjectured to induce a mixing of the early universe [61]. Rigorous results on this correspondence have been missing for 40 years. First answers are given in [10, 55, 56], albeit excluding the neighborhoods of the Taub points. Questions on the passage near the bifurcation points and the global return of trajectories are still open. Answers are required for the discussion of the Mixmaster idea and the BKL-conjecture [11] on the approach to the big-bang singularity, see also [40] for current state of the art.

Chapter 14
Application: Fluid Flow in a Planar Channel, Spatial Dynamics with Reversible Bogdanov-Takens Bifurcation

In [1, 2] the Kolmogorov problem of viscous incompressible planar fluid flow under external spatially periodic forcing has been studied. Kirchgässner reduction has been used to find time-independent bounded solutions at the onset of instability of the system when the Reynolds number increases. We regard bounded solutions as evolutions in the unbounded direction of a cross-sectional profile, and find a six-dimensional center manifold. Three conserved quantities yield a reduction to a three-dimensional reversible system with a line of equilibria. When we take the Reynolds number into account, a Bogdanov-Takens point along a one-parameter family of lines of equilibria appears, see Chap. 10. Additional reversibilities, however, change the resulting dynamics.

Consider the viscous incompressible fluid flow governed by the two-dimensional Navier-Stokes equations

$$\partial_t u = \nu \Delta u - (u \cdot \nabla)u - \tfrac{1}{\rho}\nabla p + \begin{pmatrix} \mathfrak{f}(\xi_2) \\ 0 \end{pmatrix}$$

$$0 = \nabla \cdot u \tag{14.1}$$

on the plane channel $\xi = (\xi_1, \xi_2) \in \mathbb{R} \times S^1 = \mathbb{R} \times \mathbb{R}/2\pi\mathbb{Z}$ with periodic boundary conditions. The forcing is assumed to be independent of ξ_1 and acts only in ξ_1-direction, see also Fig. 14.1.

Kolmogorov's original suggestion of a force is

$$\mathfrak{f}(\xi_2) = \sqrt{2}\sin\xi_2, \tag{14.2}$$

see also [60]. This forcing gives rise to two symmetries

$$
\begin{aligned}
S_1 &: \xi_1 \mapsto -\xi_1, \; \xi_2 \mapsto -\xi_2, \\
S_2 &: \xi_1 \mapsto -\xi_1, \; \xi_2 \mapsto \xi_2 + \pi.
\end{aligned}
\tag{14.3}
$$

© Springer International Publishing Switzerland 2015
S. Liebscher, *Bifurcation without Parameters*, Lecture Notes in Mathematics 2117,
DOI 10.1007/978-3-319-10777-6__14

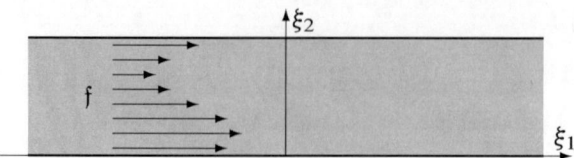

Fig. 14.1 Fluid flow in a plane channel

A less symmetric, generalized forcing

$$\mathfrak{f}(\xi_2) \; = \; c_1 \sin \xi_2 + c_2 \sin 2\xi_2 \tag{14.4}$$

breaks the second symmetry.

In both cases, the basic steady state

$$u(\xi_1, \xi_2) \; = \; (U(\xi_2), 0)^{\mathrm{T}} \tag{14.5}$$

of zero average, $\langle U \rangle = 0$, becomes unstable with increasing Reynolds number $R = v^{-2}$. The classical approach to this bifurcation through imposing artificial periodic boundary conditions in ξ_1 fails, because the onset of instability at the critical Reynolds number is due to long-wavelength instabilities [60], see also [3].

Therefore, in [1, 2] the unbounded domain is considered and stationary solutions are regarded as evolutions of a cross sectional profile $u(\xi_1, \cdot)$ evolving in ξ_1. Although the corresponding initial-value problem is ill-posed for the elliptic stationary problem (14.1), on a center manifold of the basic steady state (14.5) this *spatial dynamics* is well posed. This reduction method goes back to Kirchgässner [42, 45].

The center manifold of the ξ_1-flow turns out to be six-dimensional, with a three-dimensional set of equilibria and 3 first integrals. The level sets of the integral are, however, not transverse to the manifold of equilibria. In a critical level set, a line of equilibria survives. along this line, viewed as a one-parameter family of lines together with the Reynolds number as parameter, a Bogdanov-Takes point appears.

After suitable rescaling, the normal form, written as a third-order equation, reads

$$\dddot{y} + \dot{y} - 3y^2 \dot{y} \; = \; a y \ddot{y} + b \dot{y}^2 + \text{ small terms} \tag{14.6}$$

The parameter, i.e. the Reynolds number is already scaled out. The line of equilibria is given by $\{\dot{y} = \ddot{y} = 0\}$. Note the time-reversibilities

$$\underbrace{\begin{pmatrix} -1 & & \\ & 1 & \\ & & -1 \end{pmatrix}}_{S_1} \quad \underbrace{\begin{pmatrix} 1 & & \\ & -1 & \\ & & 1 \end{pmatrix}}_{S_2 \,(a = b = 0 \text{ only})} \tag{14.7}$$

with respect to $(\ddot{y}, \dot{y}, y)^T$ and inherited from (14.3).

The Kolmogorov forcing (14.2) gives rise to both reversibilities. In particular, the line of equilibria is then enforced by the reversibility S_2 with two-dimensional fixed-point space, see also Sect. 1.2.3. The generalized forcing (14.4) gives rise to reversibility S_1, only. The fixed-point space of this reversibility is of dimension one and cannot enforce the line of equilibria.

System (14.6) is indeed the normal form of a Bogdanov-Takens bifurcation without parameters, see Chap. 10, with additional reversibility S_1 with one-dimensional fixed-point space. Instead of a plane of equilibria we start with a one-parameter family of lines of equilibria as we have found in the Kolmogorov problem. Both settings yield the same rescaled normal form, as in Chap. 10. Given a vector field

$$\dot{z} = F(z, \lambda), \qquad z = (x, y) \in \mathbb{R}^2 \times \mathbb{R}, \quad F = (f_1, f_2, g), \quad \lambda \in \mathbb{R}, \qquad (14.8)$$

with $F(0, y, \lambda) \equiv 0$ and the linearization at the origin

$$DF(0, 0) = \begin{pmatrix} 0 & 0 & 0 \\ 1 & 0 & 0 \\ 0 & 1 & 0 \end{pmatrix}. \qquad (14.9)$$

In addition to the generic setting of Chap. 10, we assume the reversibility S_1, see (14.7),

$$F(S_1 z, \lambda) = -S_1 F(z, \lambda). \qquad (14.10)$$

Instead of a complete normal form, we use the "crude" transformation $\tilde{z} = \Psi(z, \lambda)$,

$$\begin{aligned}
\tilde{x}_1 &= D_z g(z, \lambda) \cdot F(z, \lambda) = x_1 + \cdots, \\
\tilde{x}_2 &= g(z, \lambda) &= x_2 + \cdots, \\
\tilde{y} &= y.
\end{aligned} \qquad (14.11)$$

Due to (14.10), the transformation commutes with the reversibility,

$$\begin{aligned}
\Psi(S_1 z, \lambda) &= \begin{pmatrix} D_z g(S_1 z, \lambda) \cdot F(S_1 z, \lambda) \\ g(S_1 z, \lambda) \\ -y \end{pmatrix} \\
&= \begin{pmatrix} (S_1 D_z g(z, \lambda)) \cdot (-S_1 F(z, \lambda)) \\ g(z, \lambda) \\ -y \end{pmatrix} \\
&= \begin{pmatrix} -D_z g(z, \lambda) \cdot F(z, \lambda) \\ g(z, \lambda) \\ -y \end{pmatrix} \\
&= S_1 \Psi(z, \lambda).
\end{aligned} \qquad (14.12)$$

Thus, the transformed vector field

$$
\begin{aligned}
\dot{\tilde{x}}_1 &= \tilde{f}_1(\tilde{z}, \lambda), \\
\dot{\tilde{x}}_2 &= \tilde{x}_1, \\
\dot{\tilde{y}} &= \tilde{x}_2
\end{aligned}
\tag{14.13}
$$

is again reversible under S_1. In particular, the Taylor expansion of $\tilde{f}_1$ contains only monomials of the form $\tilde{z}^\alpha = \tilde{x}_1^{\alpha_1} \tilde{x}_2^{\alpha_2} \tilde{y}^{\alpha_3}$ with $\alpha_1 + \alpha_2 > 0$ (equilibria $x = 0$), $\alpha_1 + \alpha_2$ even (reversibility S_1), and $\alpha_1 + \alpha_2 + \alpha_3 \geq 2$ (linearization (14.9)). The rescaling

$$
\begin{aligned}
\tilde{x}_1 &= \sigma^3 \hat{x}_1, \\
\tilde{x}_2 &= \sigma^2 \hat{x}_2, \\
\tilde{y} &= \sigma^1 \hat{y}, \\
\tilde{\lambda} &= \sigma^2 \hat{\lambda}, \\
t &= \sigma^{-1} \hat{t}.
\end{aligned}
\tag{14.14}
$$

then yields

$$
\begin{aligned}
\hat{x}_1' &= c_1 \hat{\lambda} \hat{x}_2 + c_2 \hat{x}_1^2 + c_3 \hat{x}_1 \hat{y} + c_4 \hat{x}_2^2 + c_5 \hat{x}_2 \hat{y}^2 + \mathcal{O}(\sigma), \\
\hat{x}_2' &= \hat{x}_1, \\
\hat{y}' &= \hat{x}_2.
\end{aligned}
\tag{14.15}
$$

We impose the following non-degeneracy conditions, $c_1 \neq 0$, $c_3 \neq 0$, to ensure a versal unfolding of the nilpotent linearization (14.9) in (y, λ), and $c_5 \neq 0$. Then coefficients can be normalized to obtain (14.6), as claimed.

14.1   Fully Symmetric Case

In the case $a = b = 0$ of both reversibilities and after dropping small terms of higher order, we find the integrable system

$$
\begin{aligned}
0 &= \ddot{y} + \dot{y} - 3y^2 \dot{y}, \\
\Theta &= \ddot{y} - y^3 + y, \\
H &= \tfrac{1}{2}\dot{y}^2 - \tfrac{1}{4}y^4 + \tfrac{1}{2}y^2 - \Theta y,
\end{aligned}
\tag{14.16}
$$

a Hamiltonian core on each level set of Θ, see Fig. 14.2a. Bounded solutions are given by a "bubble" of periodic orbits $y_{\mathrm{per}}^{\Theta, H}$ parametrized by the two conserved quantities Θ, H. The boundary of the periodic bubble consists of homoclinic orbits $y_{\mathrm{hom}}^{\Theta, H}$ and a heteroclinic pair $y_{\mathrm{het}}^{0, \frac{1}{4}}$, see Fig. 14.2b. All bounded orbits intersect a Poincaré section $\{\ddot{y} = 0\}$ parametrized by Θ, H as shown in Fig. 14.2c,d. The only exceptions are the heteroclinic orbits: here the pair has one intersection with the section.

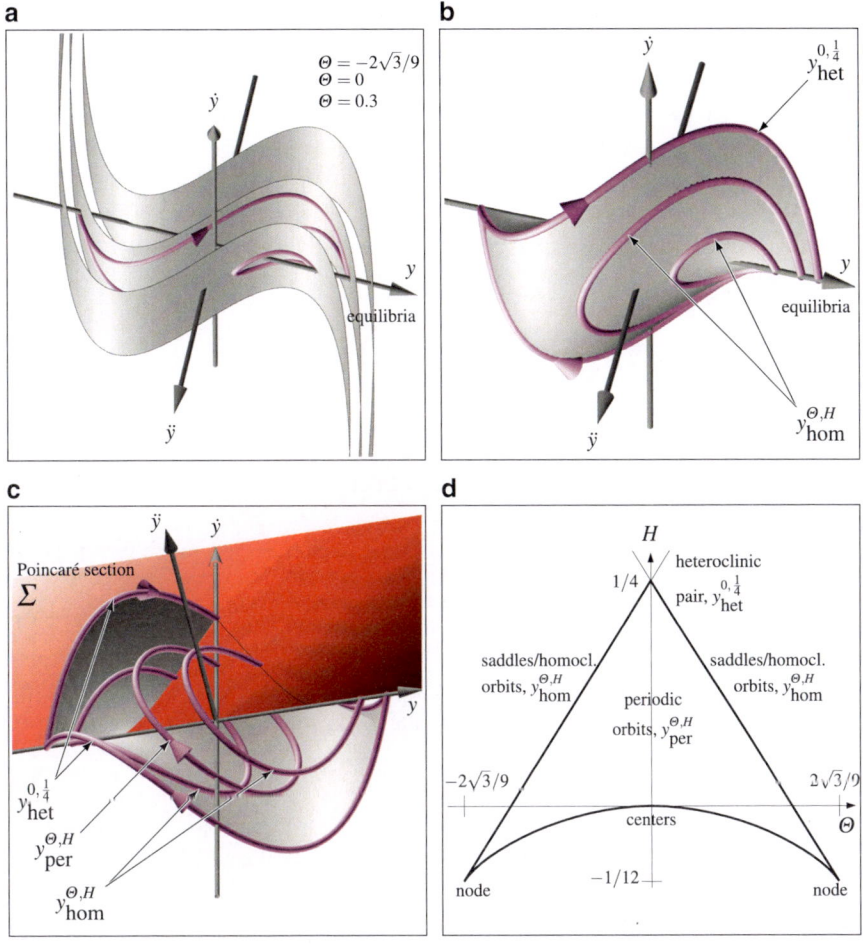

Fig. 14.2 Fully reversible Bogdanov-Takens point. (**a**) Θ-foliation and Hamiltonian core. (**b**) "Periodic bubble", set of bounded solutions. (**c**) Poincaré section. (**d**) Triangle of bounded solutions in the Poincaré section parametrized by Θ, H

Perturbations respecting both reversibilities, or at least reversibility S_2 with two-dimensional fixed-point space $\{\dot{y} = 0\}$, preserve the periodic bubble. Indeed, all periodic orbits intersect the fixed-point space twice and transversely. Transverse intersections are preserved by small perturbations, Thus the orbits of the perturbed system intersect the fixed-point space twice and, hence, are periodic.

For a thorough discussion of this case and its implications for the fluid-flow problem, see [1].

14.2 Symmetry-Breaking Perturbations

The general case $a, b \neq 0$ of (14.6) has been studied in [2]. However, a, b are assumed to be small. In other words, reversibility S_2 of the fully integrable system (14.16) is broken by a small perturbation which still respects the other reversibility S_1,

$$\ddot{y} + \dot{y} - 3y^2 \dot{y} = \varepsilon a y \ddot{y} + \varepsilon b \ddot{y}^2 + \text{ small terms} \tag{14.17}$$

The former first integrals

$$\begin{aligned}
\Theta &= \ddot{y} - y^3 + y \\
H &= \tfrac{1}{2}\dot{y}^2 - \tfrac{1}{4}y^4 + \tfrac{1}{2}y^2 - \Theta y
\end{aligned} \tag{14.18}$$

are no longer conserved but subject to a slow drift

$$\begin{aligned}
\dot{\Theta} &= \varepsilon(a\ddot{y}y + b\dot{y}^2) = \varepsilon a(\Theta - y + y^3)y + \varepsilon 2b(H - \tfrac{1}{2}y^2 + \tfrac{1}{4}y^4 + \Theta y), \\
\dot{H} &= -y\dot{\Theta}.
\end{aligned} \tag{14.19}$$

Averaging over the fast rotation inside the periodic bubble yields this drift to leading order. This drift is interpreted as a flow on the Poincaré section $\Sigma = \{\ddot{y} = 0\}$, see Fig. 14.2c.

$$\begin{aligned}
\dot{\Theta} &= \varepsilon \oint a\ddot{y}y + b\dot{y}^2 \, d\tau \quad = \varepsilon(b-a) \oint \dot{y}^2 \, d\tau, \\
\dot{H} &= \varepsilon \oint -a\ddot{y}y^2 - b\dot{y}^2 y \, d\tau = \varepsilon(2a-b) \oint \dot{y}^2 y \, d\tau.
\end{aligned} \tag{14.20}$$

Dropping the coefficient ε, the Poincaré return map of the full system is given by some first-order discretization with step size ε of the flow

$$\begin{aligned}
\dot{\Theta} &= (b-a) \oint \dot{y}^2 \, d\tau, \\
\dot{H} &= (2a-b) \oint \dot{y}^2 y \, d\tau.
\end{aligned} \tag{14.21}$$

See Fig. 14.3. Note that Θ becomes a Lyapunov function for (14.21) under the non-degeneracy condition $b \neq a$. Indeed, for $b < 1$, chosen w.l.o.g., Θ strictly decreases along solutions except at equilibria.

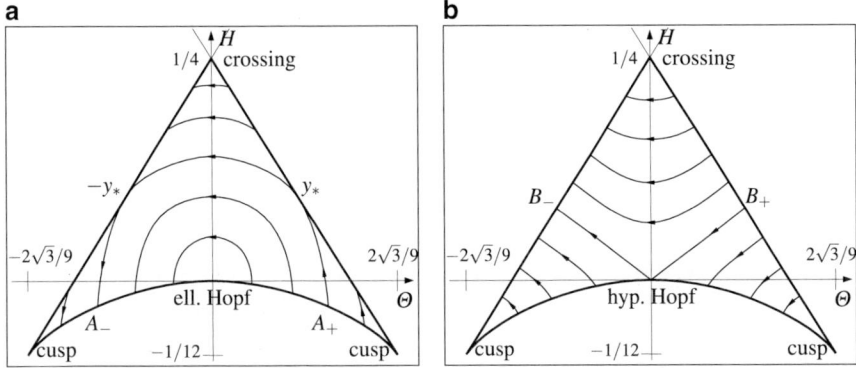

Fig. 14.3 Reversible Bogdanov-Takens point, Poincaré flow. (**a**) Elliptic Hopf point and Melnikov zeros, for $a(b-a) < 0$. *Arrows* indicate the flow direction for $a > b$, and have to be reversed in case $a < b$. (**b**) Hyperbolic Hopf point, without Melnikov zeros, for $a(b-a) > 0$. *Arrows* indicate the flow direction for $a > b$, and are reversed for $a < b$

The origin $(\Theta, H) = (0,0)$ corresponds to the origin of (14.17) and is a Poincaré-Andronov-Hopf point, see Chap. 5. Indeed, the linearization of (14.17) at equilibria $(\ddot{y}, \dot{y}, y) = (0, 0, y_c)$ reads

$$\ddot{y} + \dot{y} - 3y_c^2 \dot{y} = \varepsilon a y_c \ddot{y}. \tag{14.22}$$

It yields non-trivial eigenvalues

$$\mu_\pm = \tfrac{1}{2}\varepsilon a y_c \pm \sqrt{\tfrac{1}{4}\varepsilon^2 a^2 y_c^2 - 1 + 3y_c^2}. \tag{14.23}$$

We find a center at $y_c = 0$, a spiral sink for $a y_c \lessgtr 0$, and a spiral source for $a y_c \gtrless 0$. (For $3y_c^2 > 1$ we find saddles, the points $y_c = \pm 1/\sqrt{3}$ are the cusp points in Fig. 14.3.) Comparing the direction of the flow in Θ and the change of stability along the line of equilibria with the analysis of Chap. 5, we find an elliptic Poincaré-Andronov-Hopf point for $a(b-a) < 0$ and a hyperbolic Poincaré-Andronov-Hopf point for $a(b-a) > 0$. This also motivates an additional non-degeneracy condition

$$a(b-a) \neq 0 \tag{14.24}$$

In the hyperbolic case, we can follow the return of the strong stable/unstable manifolds of the saddles equilibria. We find regions of continuous families of saddle-focus heteroclinic connections and a region of discrete saddle-saddle heteroclinics. The boundary is given by the stable/unstable cones of the hyperbolic Hopf point, see Fig. 14.4. In the Poincaré-flow Fig. 14.3b the points $B_\pm$ denote this boundary. Saddle-saddle heteroclinic orbits in original coordinates are sketched in Fig. 14.5.

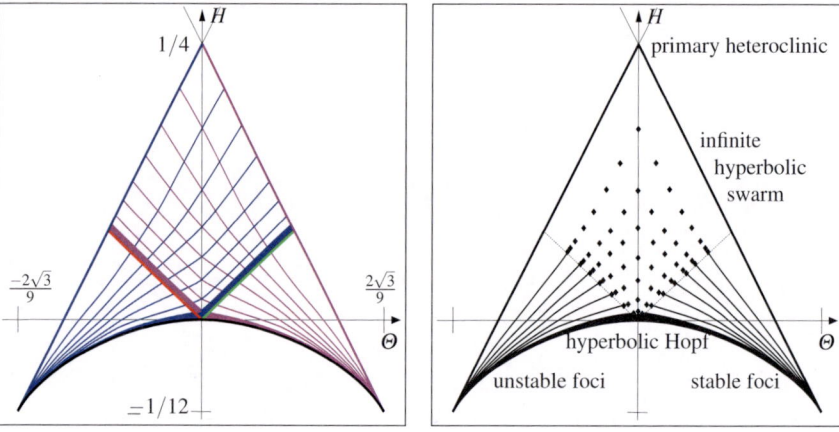

Fig. 14.4 Reversible Bogdanov-Takens point, Poincaré map, hyperbolic case. *Left side*: Simplest scenario of strong stable and strong unstable manifolds of saddles in the hyperbolic case $0 > a > b$. Angles of intersection are exaggerated. Coding: *magenta* $= W^{cu}$(saddle), *blue* $= W^{cs}$(saddle), *red* $= W^{uu}$(focus), and *green* $= W^{ss}$(focus). *Right side*: Simplest scenario of the set $\mathcal{B}_0$ of all bounded solutions in the hyperbolic case $0 > a > b$. All bounded solutions are heteroclinic

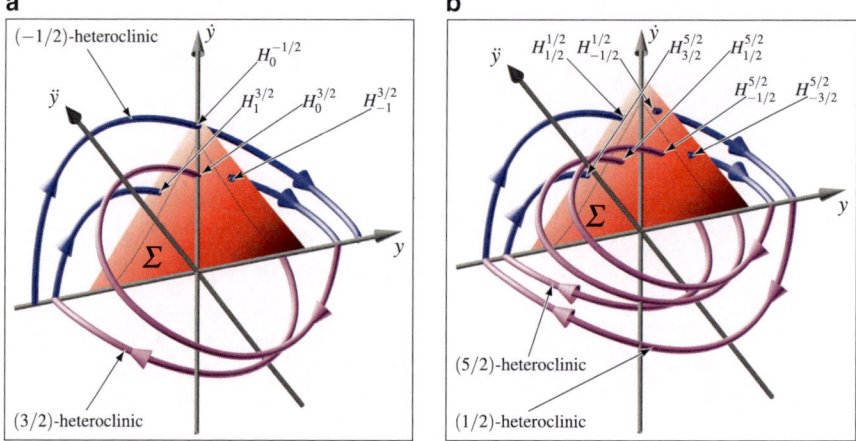

Fig. 14.5 Reversible Bogdanov-Takens point, heteroclinic orbits. Three-dimensional view of saddle-saddle heteroclinics and their intersection with the Poincaré plane. Orbits hit the symmetry line $\text{Fix}(S_1)$, the $\dot{y} - axis$, in its positive part inside the Poincaré section, (**a**), or in its negative part outside the Poincaré section, (**b**)

The elliptic case is more relevant for the Kolmogorov problem, as the normal-form reduction of the stationary PDE problem yields (14.6) with $b = 0$, hence only the elliptic case occurs.

We find a point y_* and its image $-y_*$ of tangency of the Poincaré flow (14.21) to the boundary

$$\Theta(y_c) = y_c - y_c^3, \qquad H(y_c) = 3y_c^4/4 - y_c^2/2, \qquad |y_c| > 1/\sqrt{3}, \qquad (14.25)$$

see (14.18) and Fig. 14.3a. The boundary represents the homoclinic orbits to saddle equilibria of the reversible system, $\varepsilon = 0$. The splitting of these homoclinic orbits for $\varepsilon > 0$ near y_* is determined by a Melnikov integral. This Melnikov integral is given by the angle between the Poincaré flow (14.21) and the boundary (14.25). It has a simple zero at y_* as the boundary point is varied.

This crossing is transverse [2]. We find a saddle-saddle heteroclinic orbit close to the homoclinic orbit to y_* of the integrable system. It connects to distinct equilibria close to y_*. In the Poincaré return map, Fig. 14.6, this orbit is represented by Y_0^+. Points Y_k^+ denote its iterates under the return map. The k-th return of the

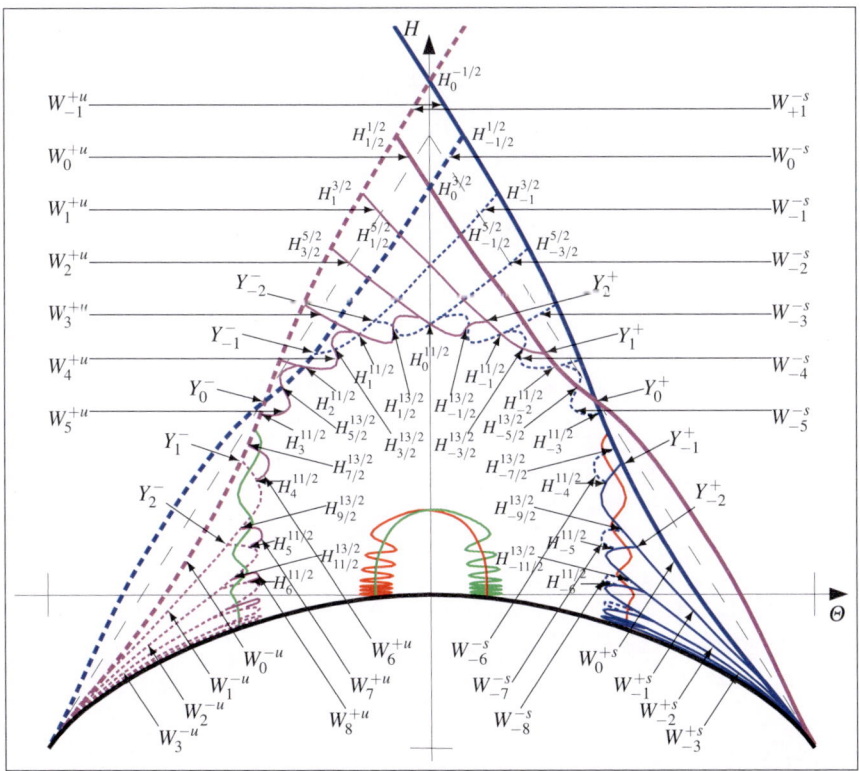

Fig. 14.6 Reversible Bogdanov-Takens point, Poincaré map, elliptic case. Simplest scenario of strong stable and strong unstable manifolds of saddles in the elliptic case $0 < a > b$. Angles of intersection are exaggerated. Coding: *magenta* $= W^{cu}$(saddle), *blue* $= W^{cs}$(saddle), *red* $= W^{uu}$(focus), and *green* $= W^{ss}$(focus)

Fig. 14.7 Set of bounded
orbits in the Poincaré section,
elliptic case. Simplest
scenario of the set $\mathscr{B}_0$ of all
bounded solutions in the
elliptic case $0 < a > b$. All
bounded solutions are
heteroclinic

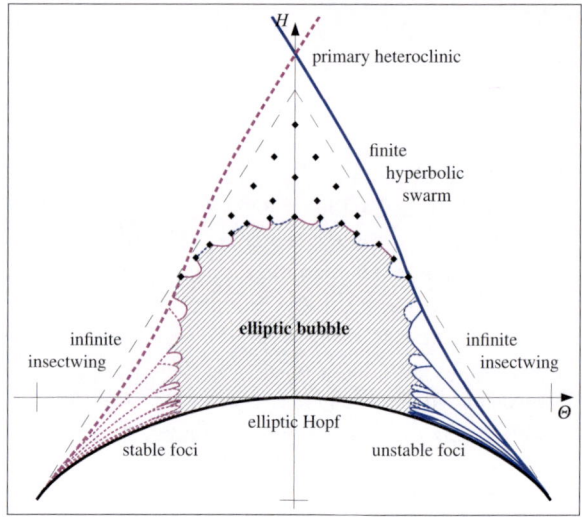

strong stable and unstable manifolds to equilibria $y_c > 1/\sqrt{3}$ is denoted by W_k^{+s}
and W_k^{+u}. Intersections of W_k^{+u} with W_ℓ^{-s}—the ℓ-th return of the strong stable
manifolds to equilibria $y_c < -1/\sqrt{3}$—are heteroclinic saddle-saddle connections
winding $k + l + 3/2$ times around the y-axis, see Fig. 14.5. The region of finitely
many saddle-saddle connections are bounded be the elliptic Hopf bubble filled with
source-sink heteroclinics and touching the lines of saddles near y_*. Close to the
cusp points $y = \pm 1/\sqrt{3}$ we find saddle-sink and source-saddle heteroclinics.

At the boundary of the elliptic Hopf bubble, separatrices split with finite angle
determined by the simple zero of the Melnikov integral at y_*.

Figure 14.7 sketches the set of bounded solutions near the Bogdanov-Takens
point, in the elliptic case. Every point of this set represents a bounded stationary
profile of the original fluid-flow problem (14.1) on a plane channel with the
generalized forcing (14.4). See [2] for further details.

Part IV
Beyond Codimension Two

Chapter 15
Codimension-One Manifolds of Equilibria

Here we discuss a special situation in which we can deal with singularities of arbitrary codimension. In Chaps. 4, 8, and after normal form transformation also in Chaps. 5, 9, we removed the manifold of equilibria by multiplying with a singular factor $1/x$ or $1/r$. This idea required that there is only one transverse direction to the manifold of equilibria. For such manifolds of codimension one, in phase space, we can generalize the idea.

We consider the general case of a manifold of equilibria of codimension one,

$$\begin{pmatrix} \dot{x} \\ \dot{y} \end{pmatrix} = F(x, y) = \begin{pmatrix} f(x, y) \\ g(x, y) \end{pmatrix}, \qquad x \in \mathbb{R}, \quad y \in \mathbb{R}^m. \tag{15.1}$$

Typically, such a system will arise as a reduced system on a center manifold of finite smoothness. Following the discussion in the Sect. 8.3, we obtain the following theorem.

Theorem 15.1 *There exists a generic subset of the class of all smooth vector fields (15.1) with an equilibrium manifold $\{x = 0\}$ of codimension one. For every vector field in that class the following holds true:*

At every point $(x = 0, y)$ the vector field is locally flow equivalent to an m-parameter family

$$\dot{z}_m = \pm z_m^{\ell+1} + \sum_{k=0}^{\ell-1} z_k z_m^k + \mathcal{O}(z_m^N), \tag{15.2}$$

$0 \leq \ell \leq m$, *of vector fields on the real line. Here N is the arbitrary but finite normal-form order bounded by the smoothness of the initial vector field (15.1), $f, g \in \mathscr{C}^M$, $N \leq M$, $N < \infty$. This is a versal unfolding of the singularity $\dot{z}_m = \pm z_m^{\ell+1}$ at the origin.*

© Springer International Publishing Switzerland 2015
S. Liebscher, *Bifurcation without Parameters*, Lecture Notes in Mathematics 2117,
DOI 10.1007/978-3-319-10777-6_15

In particular, near bifurcation points of codimension m, that appear robustly at isolated points on the equilibrium manifold, the vector field is locally flow equivalent to

$$\dot{z}_m = \pm z_m^{m+1} + \sum_{k=0}^{m-1} z_k z_m^k + \mathcal{O}(z_m^N), \tag{15.3}$$

i.e. an universal unfolding of the singularity $\dot{z}_m = \pm z_m^{m+1}$ at the origin.

Proof The equilibrium condition $f(0, y) = g(0, y) = 0$ for all $y \in \mathbb{R}^m$ allows us to factor out x.

$$F(x, y) = x\tilde{F}(x, y) = x \begin{pmatrix} \tilde{f}(x, y) \\ \tilde{g}(x, y) \end{pmatrix}. \tag{15.4}$$

The resulting vector field $\tilde{F} : \mathbb{R}^{m+1} \to \mathbb{R}^{m+1}$ does not vanish on the m-dimensional submanifold $\{x = 0\}$, for generic F. Without loss of generality, consider a neighborhood $U \subset \mathbb{R}^{m+1}$ of the origin.

We can apply the flow-box theorem to $\tilde{F}$: Take a local smooth section

$$\Sigma : \mathbb{R}^m \supset V \longrightarrow U, \tag{15.5}$$

through the origin, $\Sigma(0) = 0$, transverse to the vector field $\tilde{F}$ in U. Let $\tilde{\Phi}_t$ be the flow generated by $\tilde{F}$. Then the flow-box transformation

$$h(z_0, \ldots, z_m) = \tilde{\Phi}_{z_m}(\Sigma(z_0, \ldots, z_{m-1})) \tag{15.6}$$

transforms $\tilde{F}$ into the constant vector field $[Dh]^{-1}(\tilde{F} \circ h) = (0, \ldots, 0, 1)$. Again, $\tilde{\Phi}_t$ denotes the flow to the vector field $\tilde{F}$. Applying the transformation h to the vector field $F|_U$, we obtain an m-parameter family $[Dh]^{-1}(F \circ h) = (0, \ldots, 0, \pi_x h)$ of vector fields on the real line in a neighborhood V of the origin.

Classification of germs of vector fields and their versal unfoldings is the topic of singularity or catastrophe theory.

Singularities on the real line have the form $\dot{z}_m = \pm z_m^{\ell+1}$. In generic m-parameter families at most $m+1$ leading coefficients of the Taylor expansion vanish, i.e. $\ell \leq m$ and

$$\dot{z}_m = \pm z_m^{\ell+1} + \sum_{k=0}^{\ell-1} \zeta_k(z_0, \ldots, z_{m-1}) z_m^k + \mathcal{O}(z_m^{\ell+2}).$$

The coefficient ζ_ℓ vanishes by linear transformation of z_m. Furthermore, the map $(z_0, \ldots, z_{m-1}) \mapsto (\zeta_0, \ldots, \zeta_{\ell-1})$ has full rank, generically. Remainder terms, $\mathcal{O}(z_m^{\ell+2})$, can be pushed to any finite normal-form order, by a suitable coordinate change. This procedure yields system (15.2). See also [17], chapter 6.

Genericity conditions are expressed as algebraic conditions on the coefficients of the Taylor expansion at the origin. These conditions correspond via (15.6) to generic conditions on F.

The versal unfolding (15.2), one the other hand, is a system of the form (15.1). Therefore, it represents the versal unfolding of a generic singularity along m-dimensional manifolds of equilibria in $(m + 1)$-dimensional phase space.   □

The removal of the manifold of equilibria by a scalar, albeit singular, multiplier greatly facilitates the analysis but restricts it to the case of manifolds of codimension one in the phase space, see (8.14) and (15.4).

Most bifurcations previously discussed do not fall into this class, most notably Poincaré-Andronov-Hopf- and Bogdanov-Takens bifurcations. Their analysis uses a blow-up or rescaling procedure reminiscent of the scalar multiplier used here. It seems worthwhile to closer connect these bifurcations without parameters to singularity theory. This might provide a suitable setting to include singularities of the set of equilibria and generalize the manifold to varieties.

Chapter 16
Summary and Outlook

Along given manifolds of equilibria, bifurcations without parameters display a surprisingly rich and intricate structure of heteroclinic connections. Although manifolds of equilibria appear to be a rather degenerate feature of a vector field, the large variety of applications exhibiting this structure requires a systematic analysis of the emerging bifurcation problems. Techniques including center manifolds, normal forms and blow-up methods are indispensable for the theory.

Despite the possibility of similar algebraic classification, bifurcations without parameters are beyond the scope of classical bifurcation theory. Furthermore, dynamical properties of the respective bifurcation types differ significantly from their classical counterparts. Classical bifurcations are a degenerate case of bifurcations without parameters. In applications, it is of vital importance to check the non-degeneracy conditions of the respective bifurcations types to distinguish the generic case, without parameters, from the degenerate case, with—probably hidden—parameters.

We attempted a systematic approach towards a classification of bifurcations without parameters. Zero-Hopf and Hopf-Hopf points still require further study. Aside from the investigations of bifurcations of codimension three and beyond, several directions of future research promise interesting results.

16.1 Singularity Theory

Bifurcation theory is closely related to singularity theory, or catastrophe theory. This relation has been exploited in Chaps. 4, 8, 15 to study bifurcations without parameters along equilibrium manifolds of codimension one in phase space. In these cases, the manifold of equilibria could be desingularized by a scalar multiplier,

© Springer International Publishing Switzerland 2015   135
S. Liebscher, *Bifurcation without Parameters*, Lecture Notes in Mathematics 2117,
DOI 10.1007/978-3-319-10777-6__16

such that the resulting vector field fits into the framework of singularity theory. An extension of this approach to manifolds with more than one transverse direction is necessary.

Furthermore, the bifurcations studied here exhibit singularities of the vector field along smooth manifolds of equilibria. Singularities of the manifold itself have not been discussed. Their study will enrich the theory and provides an even closer connection of bifurcation theory and singularity theory.

16.2   Symmetries

In our classification of bifurcations without parameters of codimension one and two, symmetries only appeared as normal-form symmetries near Poincaé-Andronov-Hopf points. Applications, however, frequently exhibit additional symmetries due to their geometric properties or particular modeling assumptions. Cosmological models of Bianchi type, Chap. 13, and fluid flows in a plane channel, Chap. 14, are examples.

The methods used for the generic cases remain applicable in equivariant settings and together with classical equivariant bifurcation theory [20] should be extended to a rigorous equivariant bifurcation theory without parameters.

16.3   Global Bifurcation

The bifurcation analysis, presented here, has been local. It has been our aim to describe all solutions emerging from the bifurcations, i.e. all solutions which stay in a small neighborhood of the origin for all times. We have been rewarded with sets of heteroclinic orbits of intriguing complexity.

Aside from the study of small bounded solutions, the passage near manifolds of equilibria and their bifurcation points is a question of vast importance. Quantitative estimates of those passages are for example needed to answer relevant questions in Bianchi models, Chap. 13. Blow-up methods and rescaling methods are crucial tools here.

In singularly perturbed problems, they have been successfully used to study global trajectories which pass by singularities. A fixed rescaling of coordinates does not suffice, though. Full spherical blow ups are required. Alternatively several charts covering the blow-up sphere have to be studied to follow a passing orbit [46, 48]. Adaptation of these methods to bifurcations without parameters is necessary.

16.4   Recurrence

Contrary to classical bifurcation theory, no recurrent dynamics has been found so far near bifurcation points without parameters. Only in mixed cases of families of manifolds of equilibria we have found bifurcating equilibria or periodic orbits.

For codimension-one manifolds of equilibria discussed in Chap. 15, the drift non-degeneracy prevents any recurrent dynamics and permits a flow-box transformation of the vector field. Similar drift conditions hold true at generic Hopf and Bogdanov-Takens points. In fact, as already mentioned in the introduction, it is this drift which distinguishes bifurcations without parameters from classical bifurcations by preventing any flow-invariant transverse foliation. The cases studied so far are either "small" by codimension of the bifurcation or "small" by codimension of the manifold of equilibria in the phase space. Both leads to a strong influence of the drift conditions. The foliations of the mixed cases, on the other hand, already counteract the drift strongly enough to yield recurrent dynamics.

Recurrent dynamics might therefore be possible at bifurcation points of higher codimension: in the center manifold, the codimension of manifold of equilibria then becomes larger and the drift condition becomes less restrictive. So far, this question remains open.

Recurrence could still be induced by global properties of the systems. Such global recurrence is one of the intriguing properties of the Bianchi cosmologies introduced in Chap. 13. This is another reason to embed to the local analysis of bifurcations without parameters into global structures.

But even without recurrent dynamics, the structure of heteroclinic orbits found close to bifurcations without parameters is astonishingly rich and needs to be further investigated in order to improve the reliability of the answers to corresponding problems in applications.

References

1. Afendikov, A., Fiedler, B., Liebscher, S.: Plane Kolmogorov flows and Takens-Bogdanov bifurcation without parameters: The doubly reversible case. Asymptotic Anal. **60**(3–4), 185–211 (2008)
2. Afendikov, A., Fiedler, B., Liebscher, S.: Plane Kolmogorov flows and Takens-Bogdanov bifurcation without parameters: The singly reversible case. Asymptotic Anal. **72**(1–2), 31–76 (2011)
3. Afendikov, A., Mielke, A.: Dynamical properties of the 2d Navier-Stokes flow with Kolmogorov forcing in an infinite strip for spatially non-decaying initial data. J. Math. Fluid Mech. **7**, S51–S67 (2005)
4. Alexander, J., Auchmuty, G.: Global bifurcation of phase-locked oscillators. Arch. Ration. Mech. Anal. **93**, 253–270 (1986)
5. Alexander, J., Fiedler, B.: Global decoupling of coupled symmetric oscillators. In: Dafermos, C., Ladas, G., Papanicolaou, G. (eds.) Differential Equations, Lect. Notes Math., vol. 118. Marcel Dekker, New York (1989)
6. Anosov, D.: Roughness of geodesic flows on compact riemannian manifolds of negative curvature. Dokl. Akad. Nauk SSSR **145**, 707–709 (1962). English transl. in Sov. Math. Dokl. **3**, 1068–1069 (1962)
7. Arnol'd, V.: Geometrical methods in the theory of ordinary differential equations. Grundl. Math. Wiss., vol. 250. Springer, New York (1983)
8. Arnol'd, V.: Dynamical systems V. Bifurcation theorie and catastrophe theory. Enc. Math. Sciences, vol. 5. Springer, Berlin (1994)
9. Arnol'd, V., Gusejn-Zade, S., Varchenko, A.: Singularities of differentiable maps. Volume I: The classification of critical points, caustics and wave fronts. Monographs in Mathematics, vol. 82. Birkhäuser, Stuttgart (1985)
10. Béguin, F.: Aperiodic oscillatory asymptotic behavior for some Bianchi spacetimes. Classical Quant. Grav. **27**, 185,005 (2010)
11. Belinskii, V., Khalatnikov, I., Lifshitz, E.: Oscillatory approach to a singular point in the relativistic cosmology. Adv. Phys. **19**, 525–573 (1970)
12. Belitskii, G.: Functional equations and conjugacy of local diffeomorphisms of a finite smoothness class. Funct. Anal. Appl. **7**, 268–277 (1973)
13. Bogdanov, R.: Bifurcation of the limit cycle of a family of plane vector fields. Trudy Semin. Im. I. G. Petrovskogo **2**, 23–36 (1976)
14. Bogdanov, R.: Versal deformation of a singularity of a vector field on the plane in the case of zero eigenvalues. Trudy Semin. Im. I. G. Petrovskogo **2**, 37–65 (1976)

© Springer International Publishing Switzerland 2015
S. Liebscher, *Bifurcation without Parameters*, Lecture Notes in Mathematics 2117,
DOI 10.1007/978-3-319-10777-6

15. Bogdanov, R.: Bifurcation of the limit cycle of a family of plane vector fields. Sel. Mat. Sov. **1**, 373–387 (1981)
16. Bogdanov, R.: Versal deformation of a singularity of a vector field on the plane in the case of zero eigenvalues. Sel. Mat. Sov. **1**, 389–421 (1981)
17. Bruce, J., Giblin, P.: Curves and Singularities, 2nd edn. Cambridge University Press, Cambridge (1992)
18. Burgers, J.: Application of a model system to illustrate some points of the statistical theory of free turbulence. Nederl. Akad. Wefensh. Proc. **43**, 2–12 (1940)
19. Chillingworth, D., Sbano, L.: Bifurcation from a normally degenerate manifold. Proc. Lond. Math. Soc. (3) **101**(1), 137–178 (2010)
20. Chossat, P., Lauterbach, R.: Methods in equivariant bifurcations and dynamical systems. Advanced Series in Nonlinear Dynamics, vol. 15. World Scientific, Singapore (2000)
21. Chow, S.N., Hale, J.K.: Methods of Bifurcation Theory. Springer, New York (1982)
22. Dafermos, C.: Hyperbolic conservation laws in continuum physics. Grundlehren der Mathematischen Wissenschaften, vol. 325, 3rd edn. Springer, Berlin (2010)
23. Dumortier, F., Roussarie, R.: Geometric singular perturbation theory beyond normal hyperbolicity. In: Jones, C.K.R.T. et al. (eds.) Multiple-Time-Scale Dynamical Systems. Proceedings of the IMA Workshop, Minneapolis, 1997–1998, IMA Vol. Math. Appl., vol. 122, pp. 29–63. Springer, New York (2001)
24. Farkas, M.: ZIP bifurcation in a competition model. Nonlinear Anal. Theory Methods Appl. **8**, 1295–1309 (1984)
25. Fenichel, N.: Geometric singular perturbation theory for ordinary differential equations. J. Differ. Equat. **31**, 53–89 (1979)
26. Fiedler, B., Liebscher, S.: Generic Hopf bifurcation from lines of equilibria without parameters: II. Systems of viscous hyperbolic balance laws. SIAM J. Math. Anal. **31**(6), 1396–1404 (2000)
27. Fiedler, B., Liebscher, S.: Takens-Bogdanov bifurcations without parameters, and oscillatory shock profiles. In: Broer, H., Krauskopf, B., Vegter, G. (eds.) Global Analysis of Dynamical Systems, Festschrift dedicated to Floris Takens for his 60th birthday, pp. 211–259. IOP, Bristol (2001)
28. Fiedler, B., Liebscher, S.: Bifurcations without parameters: Some ODE and PDE examples. In: Li, T.-T., et al. (eds.) International Congress of Mathematicians, Vol. III: Invited Lectures, pp. 305–316. Higher Education Press, Beijing (2002)
29. Fiedler, B., Liebscher, S., Alexander, J.: Generic Hopf bifurcation from lines of equilibria without parameters: I. Theory. J. Differ. Equat. **167**, 16–35 (2000)
30. Fiedler, B., Liebscher, S., Alexander, J.: Generic Hopf bifurcation from lines of equilibria without parameters: III. Binary oscillations. Int. J. Bif. Chaos Appl. Sci. Eng. **10**(7), 1613–1622 (2000)
31. Fiedler, B., Scheurle, J.: Discretization of Homoclinic Orbits and Invisible Chaos, Mem. AMS, vol. 570. Amer. Math. Soc., Providence (1996)
32. Freistühler, H., Szmolyan, P.: Existence and bifurcation of viscous profiles for all intermediate magnetohydrodynamic shock waves. SIAM J. Math. Anal. **26**(1), 112–128 (1995)
33. Gelfreich, V.: A proof of the exponentially small transversality of the separatrices for the standard map. Commun. Math. Phys. **201**, 155–216 (1999)
34. Gibson, C.: Singular Points of smooth mappings. Pitman Res. Notes Math., vol. 25. Pitman, London, San Francisco, Melbourne (1979)
35. Golubitsky, M., Guillemin, V.: Stable mappings and their singularities. Grad. Texts in Math., vol. 14. Springer, New York (1973)
36. Golubitsky, M., Stewart, I.: The symmetry perspective. From equilibrium to chaos in phase space and physical space. Progress in Mathematics, vol. 200. Birkhäuser, Basel (2002)
37. Guckenheimer, J., Holmes, P.: Nonlinear oscillations, dynamical systems, and bifurcations of vector fields. Appl. Math. Sci., vol. 42. Springer, New York (1982)
38. Hale, J., Koçak, H.: Dynamics and bifurcations. Texts in Appl. Math., vol. 3. Springer, New York (1991)

39. Härterich, J., Liebscher, S.: Travelling waves in systems of hyperbolic balance laws. In: Warnecke, G. (ed.) Analysis and Numerical Methods for Conservation Laws, pp. 281–300. Springer, New York (2005)
40. Heinzle, J., Uggla, C.: Mixmaster: Fact and belief. Classical Quant. Grav. **26**(7), 075,016 (2009)
41. Hirsch, M., Pugh, C., Shub, M.: Invariant Manifolds. Lect. Notes Math., vol. 583. Springer, Berlin (1977)
42. Iooss, G., Mielke, A., Demay, Y.: Theory of the steady Ginzburg-Landau equation in hydrodynamic stability problems. Eur. J. Mech. B/Fluids **8**(3), 229–268 (1989)
43. Joseph, D.D.: Fluid dynamics of viscoelastic liquids. Appl. Math. Sci., vol. 84. Springer, New York (1990)
44. Keyfitz, B.L., Kranzer, H.C.: A system of non-strictly hyperbolic conservation laws arising in elasticity theory. Arch. Ration. Mech. Anal. **72**, 219–241 (1980)
45. Kirchgässner, K.: Wave-solutions of reversible systems and applications. J. Differ. Equat. **45**, 113–127 (1982)
46. Kosiuk, I., Szmolyan, P.: Scaling in singular perturbation problems: blowing up a relaxation oscillator. SIAM J. Appl. Dyn. Syst. **10**(4), 1307–1343 (2011)
47. Kröner, D.: Numerical Schemes for Conservation Laws. Wiley & Teubner, New York (1997)
48. Krupa, M., Szmolyan, P.: Extending geometric singular perturbation theory to nonhyperbolic points — fold and canard points in two dimensions. SIAM J. Math. Anal. **33**(2), 286–314 (2001)
49. Kurakin, L., Yudovich, V.: Bifurcation of the branching of a cycle in n-parameter family of dynamic systems with cosymmetry. Chaos **7**(3), 376–386 (1997)
50. Kurakin, L., Yudovich, V.: Branching of 2D tori off an equilibrium of a cosymmetric system (codimension-1 bifurcation). Chaos **11**(4), 780–794 (2001)
51. Kuznetsov, Y.: Elements of applied bifurcation theory. Appl. Math. Sci., vol. 112. Springer, New York (1995)
52. Liebscher, S.: Stabilität von Entkopplungsphänomenen in Systemen gekoppelter symmetrischer Oszillatoren. Diplomarbeit, Freie Universität Berlin (1997)
53. Liebscher, S.: Stable, oscillatory viscous profiles of weak, non-Lax shocks in systems of stiff balance laws. Dissertation, Freie Universität Berlin (2000)
54. Liebscher, S.: Dynamics near manifolds of equilibria of codimension one and bifurcation without parameters. Electron. J. Differ. Equat. **63**, 1–12 (2011)
55. Liebscher, S., Härterich, J., Webster, K., Georgi, M.: Ancient dynamics in Bianchi models: Approach to periodic cycles. Commun. Math. Phys. **305**, 59–83 (2011)
56. Liebscher, S., Rendall, A., Tchapnda, S.: Oscillatory singularities in bianchi models with magnetic fields (2012). Preprint
57. Manrubia, S., Mikhailov, A., Zanette, D.: Emergence of dynamical order. Synchronization Phenomena in Complex Systems. World Scientific, Singapore (2004)
58. Marchesin, D., Plohr, B., Shecter, S.: An organizing center for wave bifurcation in multiphase flow. SIAM J. Appl. Math. **57**, 1189–1215 (1997)
59. Marsden, J.E., McCracken, M.: The Hopf bifurcartion and its applications. Appl. Math. Sci., vol. 19. Springer, New York (1976)
60. Meshalkin, L., Sinai, J.: Investigation of the stability of a stationary solution of a system of equations for the movement of an incompressible viscous liquid. Prikl. Math. Mech. **25**, 1700–1705 (1961)
61. Misner, C.: Mixmaster universe. Phys. Rev. Lett. **22**, 1071–1074 (1969)
62. Murdock, J.: Normal forms and unfoldings for local dynamical systems. Monogr. in Math. Springer, New York (2003)
63. Neishtadt, A.: On the separation of motions in systems with rapidly rotating phase. J. Appl. Math. Mech. **48**, 134–139 (1984)
64. Riaza, R.: Manifolds of equilibria and bifurcations without parameters in memristive circuits. SIAM J. Appl. Math. **72**, 877–896 (2012)

65. Shoshitaishvili, A.: Bifurcations of topological type of a vector field near a singular point. Trudy Semin. Im. I. G. Petrovskogo **1**, 279–309 (1975)

66. Smale, S.: Structurally stable systems are not dense. Amer. J. Math. **88**, 491–496 (1966)

67. Smoller, J.: Shock waves and reaction-diffusion equations. Grundl. math. Wiss., vol. 258. Springer, New York (1983, 1994)

68. Smoller, J., Temple, B.: Astrophysical shock-wave solutions of the Einstein equations. Phys. Rev. D **51**, 2733–2743 (1995)

69. Takens, F.: Forced oscillations and bifurcations (1973). The Utrecht preprint, reproduced in Broer, H., Krauskopf, B., Vegter, G. (eds.), Global Analysis of Dynamical Systems, Festschrift dedicated to Floris Takens for his 60th birthday. IOP, Bristol 2001, pp. 211–259

70. Takens, F.: Singularities of vector fields. Publ. Math. Inst. Hautes Etud. Sci. **43**, 47–100 (1974)

71. Turing, A.: The chemical basis of morphogenesis. Phil. Trans. Roy. Soc. Lond. **327B**, 37–72 (1952)

72. Vanderbauwhede, A.: Centre manifolds, normal forms and elementary bifurcations. In: Kirchgraber, U., Walther, H.O. (eds.) Dynamics Reported 2, pp. 89–169. Teubner & Wiley, Stuttgart (1989)

73. Wainwright, J., Ellis, G. (eds.): Dynamical Systems in Cosmology. Cambridge University Press, Cambridge (2005)

74. Wainwright, J., Hsu, L.: A dynamical systems approach to Bianchi cosmologies: orthogonal models of a class A. Classical Quant. Grav. **6**(10), 1409–1431 (1989)

LECTURE NOTES IN MATHEMATICS 🐎 Springer

Edited by J.-M. Morel, B. Teissier; P.K. Maini

Editorial Policy (for the publication of monographs)

1. Lecture Notes aim to report new developments in all areas of mathematics and their applications - quickly, informally and at a high level. Mathematical texts analysing new developments in modelling and numerical simulation are welcome.

   Monograph manuscripts should be reasonably self-contained and rounded off. Thus they may, and often will, present not only results of the author but also related work by other people. They may be based on specialised lecture courses. Furthermore, the manuscripts should provide sufficient motivation, examples and applications. This clearly distinguishes Lecture Notes from journal articles or technical reports which normally are very concise. Articles intended for a journal but too long to be accepted by most journals, usually do not have this "lecture notes" character. For similar reasons it is unusual for doctoral theses to be accepted for the Lecture Notes series, though habilitation theses may be appropriate.

2. Manuscripts should be submitted either online at www.editorialmanager.com/lnm to Springer's mathematics editorial in Heidelberg, or to one of the series editors. In general, manuscripts will be sent out to 2 external referees for evaluation. If a decision cannot yet be reached on the basis of the first 2 reports, further referees may be contacted: The author will be informed of this. A final decision to publish can be made only on the basis of the complete manuscript, however a refereeing process leading to a preliminary decision can be based on a pre-final or incomplete manuscript. The strict minimum amount of material that will be considered should include a detailed outline describing the planned contents of each chapter, a bibliography and several sample chapters.

   Authors should be aware that incomplete or insufficiently close to final manuscripts almost always result in longer refereeing times and nevertheless unclear referees' recommendations, making further refereeing of a final draft necessary.

   Authors should also be aware that parallel submission of their manuscript to another publisher while under consideration for LNM will in general lead to immediate rejection.

3. Manuscripts should in general be submitted in English. Final manuscripts should contain at least 100 pages of mathematical text and should always include

   - a table of contents;
   - an informative introduction, with adequate motivation and perhaps some historical remarks: it should be accessible to a reader not intimately familiar with the topic treated;
   - a subject index: as a rule this is genuinely helpful for the reader.

   For evaluation purposes, manuscripts may be submitted in print or electronic form (print form is still preferred by most referees), in the latter case preferably as pdf- or zipped ps-files. Lecture Notes volumes are, as a rule, printed digitally from the authors' files. To ensure best results, authors are asked to use the LaTeX2e style files available from Springer's web-server at:

   ftp://ftp.springer.de/pub/tex/latex/svmonot1/ (for monographs) and
   ftp://ftp.springer.de/pub/tex/latex/svmultt1/ (for summer schools/tutorials).

Additional technical instructions, if necessary, are available on request from lnm@springer.com.

4. Careful preparation of the manuscripts will help keep production time short besides ensuring satisfactory appearance of the finished book in print and online. After acceptance of the manuscript authors will be asked to prepare the final LaTeX source files and also the corresponding dvi-, pdf- or zipped ps-file. The LaTeX source files are essential for producing the full-text online version of the book (see http://www.springerlink.com/openurl.asp?genre=journal&issn=0075-8434 for the existing online volumes of LNM). The actual production of a Lecture Notes volume takes approximately 12 weeks.

5. Authors receive a total of 50 free copies of their volume, but no royalties. They are entitled to a discount of 33.3 % on the price of Springer books purchased for their personal use, if ordering directly from Springer.

6. Commitment to publish is made by letter of intent rather than by signing a formal contract. Springer-Verlag secures the copyright for each volume. Authors are free to reuse material contained in their LNM volumes in later publications: a brief written (or e-mail) request for formal permission is sufficient.

Addresses:
Professor J.-M. Morel, CMLA,
École Normale Supérieure de Cachan,
61 Avenue du Président Wilson, 94235 Cachan Cedex, France
E-mail: morel@cmla.ens-cachan.fr

Professor B. Teissier, Institut Mathématique de Jussieu,
UMR 7586 du CNRS, Équipe "Géométrie et Dynamique",
175 rue du Chevaleret
75013 Paris, France
E-mail: teissier@math.jussieu.fr

For the "Mathematical Biosciences Subseries" of LNM:

Professor P. K. Maini, Center for Mathematical Biology,
Mathematical Institute, 24-29 St Giles,
Oxford OX1 3LP, UK
E-mail: maini@maths.ox.ac.uk

Springer, Mathematics Editorial, Tiergartenstr. 17,
69121 Heidelberg, Germany,
Tel.: +49 (6221) 4876-8259

Fax: +49 (6221) 4876-8259
E-mail: lnm@springer.com